HOW NAZI GERMANY HAS
CONTROLLED BUSINESS

AMS PRESS
NEW YORK

THE BROOKINGS INSTITUTION

The Brookings Institution—Devoted to Public Service through Research and Training in the Social Sciences—was incorporated on December 8, 1927. Broadly stated, the Institution has two primary purposes: the first is to aid constructively in the development of sound national policies; and the second is to offer training of a super-graduate character to students of the social sciences. The Institution will maintain a series of co-operating institutes, equipped to carry out comprehensive and inter-related research projects.

The responsibility for the final determination of the Institution's policies and its program of work and for the administration of its endowment is vested in a self-perpetuating board of trustees. It is the function of the trustees to make possible the conduct of scientific research under the most favorable conditions, and to safeguard the independence of the research staff in the pursuit of their studies and in the publication of the results of such studies. It is not a part of their function to determine, control, or influence the conduct of particular investigations or the conclusions reached; but only to approve the principal fields of investigation to which the available funds are to be allocated, and to satisfy themselves with reference to the intellectual competence and scientific integrity of the staff. Major responsibility for "formulating general policies and co-ordinating the activities of the various divisions of the Institution" is vested in the president. The by-laws provide also that "there shall be an advisory council selected by the president from among the scientific staff of the Institution and representing the different divisions of the Institution."

HOW NAZI GERMANY HAS CONTROLLED BUSINESS

BY

L. HAMBURGER

WASHINGTON, D.C.

THE BROOKINGS INSTITUTION

1943

Library of Congress Cataloging in Publication Data

Hamburger, L. (Ludwig), 1901–
How Nazi Germany has controlled business.

Reprint. Originally published: Washington, D.C.:
Brookings Institution, 1943.
1. Industry and state—Germany—History.   I. Title.
HD3616.G4H3   1981   338.943′009′043   81-12769
ISBN 0-404-16939-2   AACR2

Prefatory Note

The major object of this study is to make available to the American reader a concise running story of Nazi business control. This has necessitated the omission of many details which, though interesting in themselves, might have made it difficult for the reader to see the woods because of the trees.

Preparation of such a study at a distance more than 3000 miles from Berlin presents obvious problems. Since the war the flow of German source materials has not been as full or as regular as before; and, while the author has gathered valuable first-hand information from persons who had experience on the spot—American officials, businessmen, and other observers—it has not been possible to interview any of the men charged with the control of the German economy. Some parts of the narrative are therefore not as well rounded or as close to the ground as they could have been made under normal circumstances.

In the development of this study the author has enjoyed the advantage of numerous discussions and consultations with Mr. Richard Eldridge of the Department of State. Mr. Eldridge's broad knowledge and penetrating judgment of the European economic scene, together with his advice and encouragement, have been invaluable. For background information and a number of helpful suggestions the author is indebted to Professor F. K. Mann of the American University. Source materials available in the European unit, Bureau of Foreign and Domestic Commerce, U. S. Department of Commerce, have been used extensively. For their many courtesies as well as for information generously supplied, the author wishes to thank Dr. Louis Domeratzky and Mr. H. Arnold Quirin, chief and economic analyst, respectively, of this unit. The assistance of Mrs. Irma K. Chapman, secretary of the unit, is also gratefully acknowledged. It goes without saying that the author is solely responsible for all statements of fact and all interpretations offered.

L. H.

Contents

It is an axiom of statesmanship which the successful founders of tyranny have understood and acted upon—that great changes can best be brought about under old forms.

—HENRY GEORGE, *Progress and Poverty*, Bk. 8, Ch. 2.

O N JANUARY 30, 1933 President Paul von Hindenburg appointed Adolf Hitler to the chancellorship of the German Reich. Within a few months Hitler acquired complete control of the machinery of government. His hold was strengthened as the German National-Socialist Workers [Nazi] Party gradually pervaded the whole public service.

The new government turned almost instantly to the control of economic activities. Not all fields were tackled at once, but essential spadework was done to ensure the effectiveness of more extensive action. In the ensuing six years government control reached out to every phase of economic life.

When, on September 1, 1939, the German armies invaded Poland, the pattern of government control of economic life had been practically completed; additions required by actual war have followed the lines laid down previously. Leading Nazis have taken pride in the smoothness with which this transition was effected. As they freely said, all they had to do was to round out and intensify the controls already in operation.

This rapid extension of government control over economic life was dictated by the apparent requirements of a vast mobilization program, but it stemmed also from the underlying philosophy of national socialism. This philosophy is one of regimentation. Its fundamental tenets are: that it is bad for man to be his own master; that the essential choices should be made for him; that compulsion is the integrating force of society; that freedom is an evil. Accordingly, authority replaces reason as the source of truth; obedience replaces the responsible exercise of independent judgment as the principal civic virtue; command supplants agreement and majority decision as a basic social technique; subordination supplants liberty as a national ideal. The pursuit of happiness is superseded by the attainment of order. Blind faith is acclaimed, while doubt is scorned. The individual is dwarfed; as a citizen he has only duties, no rights. Government is extolled; it wields dictatorial power in the name of the nation, the secularized God of an atheistic religion. The state is total.

Under this philosophy the Nazis, through their government, have regimented business, labor, the professions, the schools, the churches, the press, broadcasting—everybody down to the blind and the deaf mute. In this study we are concerned with only one object of Nazi regimentation—German business.

1

I. Alternative Methods of Control

The issue for the Nazis has never been *whether* they should control business. Their problem has been *how* to control it. Three different methods were open. The Nazis might have *nationalized* private enterprise. If this seemed an impracticable or too revolutionary method, some form of *commandeering* might have been adopted. Finally, there was the possibility of exercising the requisite control by means of an extensive system of *regulation*.

NATIONALIZATION

Nationalization under a scheme of government ownership might have seemed the most natural of the three methods. Certainly the stage was well set for widespread nationalization. When Hitler came into power, the Reich already had important business interests. Even prior to 1931 it owned and operated the mails, the telephone, the entire network of broadcasting stations, the telegraph, and almost all the railroads. It also had a monopoly (1) of the importation of and wholesale trade in liquor; (2) of wholesale trade in and importation and exportation of matches; and (3) of importation of corn. Moreover, the Reich had gradually acquired important interests in nitrogen, power, airlines, shipping, shipbuilding, and metal manufacturing, and it controlled almost the entire production of aluminum. Most of the latter interests were pooled in a holding company usually referred to as VIAG, which was capitalized in 1931 at 180 million reichsmarks, and was entirely Reich controlled. In the field of banking, one of the five great banks, the Reichskreditgesellschaft, was owned by the state.

In the great economic crisis of 1931-32, Chancellor Heinrich Bruening made further substantial additions to the Reich domain. Following the credit collapse, the government rapidly extended its power over the principal banks to a degree which caused a leading German economic weekly rightly to assert: "On February 22, 1932, . . . German banking was nationalized. The continued existence of a 'private sector' does not alter the validity of this statement."[1] Beginning in May 1932, the Reich swiftly achieved domination of 40 per cent of total German steel production including a large number of related interests and ramifications.[2]

---

[1] Gustav Stolper, *Der Deutsche Volkswirt*, Feb. 26, 1932, p. 703.

[2] In addition to the Reich, the German states—mainly Prussia—had substantial interests in business enterprises. These included government banks, mining and power properties, some manufacturing interests, and nearly half the German forest domain. The municipalities were also actively engaged in many lines besides public utilities.

If the Reich had taken over what remained of private banking, all the heavy industries and many light ones, mines, shipbuilding, insurance, perhaps big agricultural estates, and other key activities, an existing trend would merely have been accentuated. Such an extension of the state's ownership would not by German standards have been startling.

Nationalization of business was not favored because of both political and economic drawbacks.

Nationalization could have been accomplished either by purchase or by confiscation. Outright purchase would have involved the Nazi government in very heavy financial responsibilities and troublesome problems. For these reasons, purchase was not even contemplated. But if the government had resorted to widespread confiscation of property rights without compensation, it would have been clear that the Nazis were heading toward radical forms of socialism or even communism.

Such a course probably would not have caused the *majority of Germans* much worry. There was a long and strong tradition of socialistic thinking in Germany. Furthermore, big business had become very unpopular, especially during the years of the great depression. The Nazis had accentuated these trends by giving their program a socialistic slant through extensive use of anti-capitalistic catchwords and slogans. Actually the Nazi rank and file, articulate through Captain Roehm and his friends, were widely in favor of "taking over"; business felt decidedly insecure during the first sixteen months of the Nazi regime. When eventually Hitler decided upon a different course, declared "the revolution ended," and had the Roehm clique killed (June 30, 1934), it was neither because the German masses clung to the idea of private ownership of business enterprise nor because Hitler had promised them to fight communism. It would not have been difficult for the Nazis to "sell" state ownership of enterprise to the great majority of the German people as a basic German idea—by dressing it up in a brown shirt, branding it with the swastika, and advertising it in Nazi lingo.

On the other hand, among the German property-owning classes—which had been largely instrumental in putting Hitler in power—a policy of nationalization would have aroused the most bitter resentment. This reaction would have been shared, though to a minor degree, by not a few members of the higher civil service and the military class whose co-operation in the militarization effort was essential. In view of the hearty support they would have received from the working population, the Nazis conceivably might have disregarded opposition from these groups. But there was also an *international angle* to the problem. The Nazis had widely advertised themselves as the bulwark against communism, not only in Germany but also throughout the rest of the world. If they had taken the radical approach they would have disappointed their numerous friends abroad. Specifically, they would have alienated the sympathies of influential circles in various countries, whose attitudes and actions were to be so largely responsible for the long series of Nazi diplomatic successes. Such a course would have nipped "appeasement" in the bud. Moreover, an outright radical policy would have deterred foreign capitalists from investing in Germany after 1933.

Finally, it may well be presumed that the Nazis did not think it economically wise to adopt a policy of widespread nationalization. For, does nationalization necessarily involve or lead to effective government control of nationalized enterprise? Acquisition by the state of the ownership of business is one thing; successful operation of state-owned business is another. Soviet experience had shown that abrupt and wholesale nationalization raises management problems of the greatest magnitude. Indeed, the bulk of German state-owned enterprise acquired before 1931 required many years to build up. The Nazis did not have the time. They wanted to prepare for war. They could not indulge in long-range social experimentation. They had to make the most out of existing talent and experience.

For reasons such as these the Nazis did not go in for nationalization of private business on a broad scale. It is true they confiscated the property of enemies of the regime and of Jews—down to the last article of consumption. Confiscated businesses, however, were liquidated or turned over to former competitors and to tested Nazis, or disposed of to various private interests as in the case of corporation stock.[3] Similarly, in Austria, in Czechoslovakia, and in the defeated countries, important business and property rights were transferred to German banks, industrial concerns, commercial firms, farmers, and others—rather than to the German state. By and large property was shifted from private hands to other private hands. Economically speaking, this action meant annihilation or impoverishment of some and enrichment of others. Put bluntly, it was looting of business—not nationalizing.

[3] The only apparent exception was the shares Fritz Thyssen held in the German United Steel Corporation which the government retained. A trustee was appointed to hold these shares. See Kurt Lachmann, "More on the Hermann Goering Works," *Social Research*, September 1942, p. 399.

The nationalization method of control
was used in exceptional cases.

The Nazis resorted to nationalization in only one major case. In 1934 ownership of the Junkers Works—manufacturing the widely known airplanes—was transferred to the Reich. Incidentally, the distinguished engineer for whom the works are named was not popular with the Nazi regime. In addition, the Nazis relied on nationalization in certain minor cases. Thus, at the end of 1937 the Reich railroads took over the Brunswick Railroad Corporation, and also the Luebeck-Buechen Railroad Corporation. A majority of the stock of both companies had previously been held by the Reich; minority stockholders were compensated. Similarly, about the middle of 1938 the Reich railroads acquired ownership of the Lokal Bahn Corporation of Munich which operates the Isartal Railroad, so well known to American sightseers, and in March 1943, the Schipkau-Finsterwalde Railroad.[4]

A somewhat more significant trend toward nationalization developed in connection with the importation of agricultural products. This development grew out of the Reich's monopoly of corn imports which had been operating under the act of March 26, 1930. The Reich Corn Board's field of action was extended by an act of May 30, 1933 to include every kind of grain and feed. In the ensuing three years four additional Commodity Control Boards (Reichsstellen) were set up to deal with the importation of dairy products, vegetable and animal oils, fats, eggs, livestock, and meat products. The various boards, however, were not the exclusive importers. A good many, if not the majority, of imports continued in the hands of private business. The private importer, of course, is not a free operator. He is under obligation to offer his goods for sale to the proper Commodity Control Board, which may or may not accept them. When it does accept, it resells to the private importer at the prevailing domestic market price. If it does not accept, the goods will not obtain customs clearance and therefore are not available for sale in the home market. While the government thus controls private agricultural imports, it is clear that this segment of the German economy has been nationalized only in part.

There have been numerous additions to state-owned enter-
prise that do not represent acts of nationalization.

While instances of outright nationalization have been few and far between, it should not be assumed that the situation with respect to

[4] In addition, in some cases the Nazis expanded the business sector owned by the federal states and the municipalities.

state ownership has been frozen since the beginning of the Nazi regime. Actually, there have been important additions to the properties held by the government.

The Nazis have pursued a policy of business expansion. Many new economic activities were begun and many old ones have been greatly developed in Germany, mainly in connection with the rearmament and war effort. Some of these activities were earmarked for and have been carried on by or under the government. Among them are the construction and operation of several thousand miles of first-class highways, originally linked to the Reich railroads but later vested in a separate government authority, and the mass manufacture of an inexpensive so-called "People's Car,"[5] by an agency of the Reichs Labor Front, which also greatly expanded the banking and publishing business it had taken over from the free labor unions. In addition, the Labor Front acquired and operates hotels, pleasure ships, bathing beaches, and other recreation facilities, as well as brickmaking and construction interests.

Additions to private enterprise have been much greater than those to public enterprise.

To obtain a proper perspective, additions to state-owned enterprise should be compared with the growth in private enterprise. The latter has been enormous, both as to development of new industries and as to expansion of established industries. Prominent among the new industries are those producing synthetic raw materials such as rubber, gasoline, staple fiber (cotton, wool, silk), and resin. It is true that businessmen did not necessarily take on the new lines of their own free will; nor did they always assume the risks involved. Nonetheless, the industries were built, are owned, and are operated by private enterprise.

Of the old industries, those showing marked expansion include the manufacture of arms and ammunition, of chemicals, and of machines; shipbuilding, whaling, building construction, and related activities. Expansion has also been carried out mainly under private ownership and operations. Developments in these industries, old and new, have required billions of reichsmarks in investment funds.

Far from going in for nationalization as a policy, the Nazis have actually returned substantial portions of state-held property to private ownership. As early as 1935 the Reich transferred to the Deutsche

---

[5] None of these have been delivered to the public; eventually army vehicles, probably tanks, were made.

Bank und Diskontogesellschaft shares with a par value of 14 million reichsmarks out of a portfolio of 50 millions it owned in the stock of that institution. During the next two years the remaining 36 millions were sold and by March 1937 the "DD-Bank" was again entirely in private hands. In 1936 and 1937 the Reich followed a similar course with respect to the Commerz and Privat Bank, with the result that by September 1937 the entire capital stock of 80 million reichsmarks was again privately owned. Even more important was the action taken in the fall of 1937 when the Reich relinquished the ownership of the Dresdner Bank. Shares sold were reported to "exceed 100 million reichsmarks." In actual fact, 200 millions must have been involved, for that was the amount underwritten by the Reich in 1932. Through these transactions the Nazis gave up the dominating position in banking the Reich had acquired in 1931 and 1932. Banking, which had been practically nationalized by anti-socialist Heinrich Bruening, was denationalized by anti-plutocrat Adolf Hitler.

The Nazis also divested the Reich of ownership in certain industries. In March 1936 a controlling stock interest (3.6 million shares) in the German Shipbuilding and Engineering Company (Deschimag) was sold to a group of Bremen merchants. In September 1936 the Reich government disposed of 8.2 million shares (almost the entire capital stock) of the Hamburg South American Shipping Company. Most significant of all, in March and April 1936 the Reich completely restored to private hands the key position in the German United Steel, which had been acquired by Chancellor Bruening. In September 1941 its controlling interests in the Hamburg-American Line and in the North German Lloyd were sold to a group of business men in Hamburg and Bremen. Finally, in the summer of 1942 the Reich transferred a majority interest in the 14 million stock of the Kontropa Corporation in Vienna to a group of private bankers.[6] These transactions were business deals—not gifts. In the spring of 1941 a competent observer estimated the total yield for the Reich at no less than 1.5 billion reichsmarks.

The transfer of state-owned properties to private ownership has at least offset the additions to state ownership. When it is also considered that the lion's share of the expansion of economic activities under the Nazis was reserved for private business, it seems clear that the relative share of the state in total ownership of business has not

<hr>

[6] *Deutsche Bergwerkszeitung,* July 2, 1942. Further acts of "reprivatization" seem to be contemplated: "Verschaerfte Kaufkraftabschoepfung," *Frankfurter Zeitung,* Feb. 14, 1943.

increased. On the contrary, available evidence indicates that in relation to privately-owned properties public ownership under the Nazis has declined.

The Hermann Goering Works are not primarily a public enterprise.

The establishment and development of the Reichs Works Hermann Goering Corporation may seem an exception to the general principle of declining public ownership. These works were set up by Marshal (then General) Hermann Goering pursuant to a decree dated July 23, 1937, and were originally intended to undertake the mining and smelting of low-grade iron ore found in deposits formerly unused. The Reich government underwrote the original capital stock of 5 million reichsmarks, and a number of private corporations were requested to cede existing mining rights[7] to the Goering Works in exchange for stock of the Corporation. In April 1938 the capital of the Works was increased by 395 million reichsmarks. Of this amount, the Reich underwrote 240 millions of common stock. Of the balance, 130 millions was issued as non-voting preferred stock. An undisclosed number of iron smelting and manufacturing enterprises were forced to purchase this preferred stock, each participating firm being asked to underwrite 50 reichsmarks for each worker employed. When the Corporation was later reorganized its capital stock was increased to 652 million reichsmarks.[8] The Hermann Goering Works were thus largely owned and entirely operated by the Reich—only one representative of private business was allowed to sit on the board of directors.[9]

But what is the *raison d'être* of the Hermann Goering Works? Was the corporation really set up to ensure that low-grade iron ore would be mined and smelted? There was no need for the government to embark upon such activities. The German mining and smelting industry was willing and ready to do the job, with the usual government support; in fact, it had already made the necessary plans. In an unusual

---

[7] The acquisition of mining rights was merely a first step. In the spring of 1938, the Goering Works began to reach far beyond their original scope. Subsidiary industries were then acquired and later high grade iron ore mines, coal mines, steel plants, manufacturing industries, and shipping companies were added. Building construction and commercial activities were also undertaken. Eventually the Goering Works took part in the looting of Europe, swallowing up important business interests in Austria, in Czechoslovakia, in Rumania, in Norway, in Poland and in Lorraine. For a more complete account see Kurt Lachmann, "The Hermann Goering Works," and "More on the Hermann Goering Works," *Social Research*, February 1941, pp. 29 ff., and September 1942, pp. 396 ff.

[8] *New York Times*, Jan. 18, 1941.

[9] Hans Ilau, "Die Reichswerke Hermann Goering," *Wirtschaftskurve* (Frankfurter Zeitung) I, (1938), p. 21.

upsurge of criticism it was even openly admitted that the creation of the Works had come as a "disagreeable surprise" to private industry.[10] Nor was the corporation expanded subsequently so the Reich could ensure control over a unified, organic sector of production. The growth of the Works has been largely haphazard and disjointed. They have developed into a motley conglomeration of diversified enterprises. One is reminded of the industrial hodgepodge concocted by the ill-famed Hugo Stinnes in the wake of the German inflation of the early 1920's.

Judging only from German comment, there is something "fishy" about government ownership of the Works. As a rule, comment is either bewildered or embarrassed. Apparently German writers cannot understand, or will not truthfully explain, what is the place and function of the Works in the pattern of the Nazi economy. Obviously, they are upset by or have something to conceal in regard to a development that, without apparent necessity or logical objective, seems to reverse—in its limited field—Nazi policies in regard to ownership of business enterprise.

It would seem that state ownership is a secondary feature of the Goering Works. It is a means to an end other than government control of the plant involved. It serves ulterior motives. The Goering Corporation should be seen as a personal monument for Mr. Goering, rather than as a public enterprise. It should be seen as a business principality, a fief for the first vassal of the ruler of the Third Reich, the Nazi equivalent of what the distinguished German engineer-economist Walter Rathenau once called an industrial dukedom.

COMMANDEERING

A second method open to the Nazis was commandeering of business. This method of control had been used in the United States as early as the First World War when the government "took over" the railroads, express companies, and telegraph and telephone companies. For the Nazi government to run the plants without assuming ownership had a definite advantage as compared with nationalization: it would have seemed a far less revolutionary method.

Neither government operation with ownership unchanged nor roundabout commandeering was tried extensively.

The Nazis apparently did not consider government operation of privately-owned plants any more seriously than they did government

[10] The same, p. 17.

ownership. True, on March 23, 1937, Marshal Goering—in his capacity as Economic High Commander (see p. 15)—issued a decree providing for the appointment by the government of trustees to operate idle farms, or such farms as "were not operated in a manner that was up to the standards required to ensure adequate food supplies for the people." But it is doubtful that trustees were actually appointed, except perhaps in a few isolated cases; presumably the threat held out by the government was sufficient to spur lax or indifferent farmers. Nor, as far as we know, has the Nazi government taken over—or even threatened to take over—operation of plants other than farms.

Neither have the Nazis resorted to what one might call roundabout commandeering. It had long been a tradition in Germany for the government or municipalities—in addition to using other more indirect devices—to appoint officials to sit on the boards of many so-called mixed public and private enterprises; usually these officials were given larger powers than the representatives of private capital. It would have been natural for the Nazis to have fallen in line with this tradition. That is to say, they could have appointed government officials to sit with the usual elected members on the boards of private corporations, at least in the case of the more important banking and industrial concerns. Except in the cases of some of the new corporations, however, such action was not taken.

While leading Nazi government and party officials are known to have been given jobs in important private corporations, either as executives or as board members, the appointments were made by the corporations in an attempt to obtain favors from the government, rather than by the government with a view to influencing the operations of the corporations.[11] The object of these appointments was to evade rather than to ensure government control of business (see pp. 94-95).

Commandeering as an alternative was completely rejected by the Nazis.

REGULATION

It is clear that the Nazis have not converted business into a government department. On the contrary, they have upheld the structure of private enterprise in both its property-owning and its property-managing aspects. But the Nazis have bent private enterprise to government purposes.

---

[11] It should be noted that in 1933 and 1934 many local and regional Nazi officials brought pressure on boards of directors or officers. While they intimidated management, they did not aim at plant operation.

*The Nazis chose to control busi-
ness through regulation.*

The businessman was left in charge so the government might avail itself of his experience and talent. He has, however, been subjected to drastic restrictions and has been told how to run his firm. Rather than assuming the burden of actual ownership and operation, government under the Nazi regime has curbed and directed the conduct of private enterprise. To control business the Nazis have used the method of regulation.

On the surface, this was a conservative and relatively innocuous method of controlling business. Actually, the system of private enterprise has undergone a revolution. Under the extreme regulations that have emerged and that will be outlined in later pages, business finds itself in a position not essentially different from what it would have been if the Nazis had adopted a policy of outright nationalization or commandeering. But before describing the manner and techniques employed to regulate business, and the extent of regulation, it is necessary to describe *how* the controls evolved and *who* among the Nazi bureaucracy did the regulating.

II. Development and Co-ordination of Controls

There has been a widespread impression, particularly outside Germany, that comprehensive control of business was fully worked out in advance as part of a broad economic plan. In fact, however, there is no convincing evidence that the Nazis ever had a grandiose economic blueprint with which to start out to control business. It is true that at the beginning of February 1933—immediately following their appointment to office by von Hindenburg—Hitler and von Papen issued a manifesto proclaiming two "plans" to end the farmers' plight and the unemployment among wage earners. But this statement consisted only of generalities. It announced an economic and political goal —it was a promise, nothing more. Nazi publicity instantly dropped any current reference to "plans," and for years, any suggestion of planning was attacked as smacking of "bolshevism."[12]

Thanks to the rearmament effort and activities incident thereto, the goal set in 1933 was actually in sight by the fall of 1936. Then the Nazis could point to actual results. There was no longer any risk in claiming they had acted according to a plan. With the obvious intention of creating the impression that he had acted with wise forethought of detail, thereby magnifying the abilities of the government in the eyes of the people, Hitler told the surprised German nation in September 1936 that for almost four years it had lived under and contributed to what was now termed the "first four-year plan." This was a blueprint in retrospect; it had never existed in fact.

Nor was the "second four-year plan" anything more than a slogan to "sell" the German public a variety of measures designed to expand and speed up the militarization effort. Nazi objectives in this second period, though ambitious, were more limited than those of the successive Soviet five-year plans that the phrase immediately brings to mind. The Nazi government proposed to promote the mining and smelting of low-grade iron ore (Hermann Goering Works), and to increase further the production of synthetic and substitute raw materials and foodstuffs, as well as of arms and ammunition. While this program eventually led to co-ordination of the various economic controls then existing and to be established later, there is no evidence that a blueprint had been drawn up or carried out. Characteristically, many activities were continued and advertised as four-year plan activities after

---

[12] Perhaps the sole exception related to Dr. Schacht's "New Plan"—announced in the fall of 1934—dealing with the limited field of import and related controls.

1940 when the second four-year plan had run its course. But it has never been claimed that there was a third four-year plan.

Development of business controls was for the most part very unsystematic.

Since there was no plan *originally,* establishment of an agency of control tending towards—or actually claiming—an all-inclusive economic jurisdiction was not required at the outset. Nor have the Nazis created such an institution in later years. At no time has there been any single government office, board, bureau, or administration corresponding to the Gosplan in the Soviet Union, and charged with controlling every phase of economic life through direct or indirect orders to hundreds and thousands of factories, shipyards, mines, banks, farms, services, and the like.

Rather than setting up one all-inclusive agency, the Nazis used a motley variety of operating agencies, each for a different purpose. This was the true reflection in the institutional field of the Nazi way of attacking the job of controlling business. Their action, especially in the beginning, was empirical and bordered on the opportunistic; it was more often haphazard than systematic, disjointed rather than integrated. They did not start regulating the several phases of business life and operation simultaneously. Nor did they, except for the flow of capital, complete the regulation of each phase in one stroke. Only in the case of agriculture was practically every phase placed under regulation within two years. Steadily keeping in mind their broad rearmament and militarization objective, the Nazis have acted mainly on a "control-as-you-go" basis. Progress toward more complete control was made by slowly advancing steps, many of which were dictated by the sudden emergence of problems, difficulties, bottlenecks, rather than by deliberate action.

Evolving policies of control in piecemeal fashion, the agencies of control also evolved piecemeal. To some extent the Nazis relied on agencies existing in 1933; others they established and strengthened as need arose. Since these needs were numerous and the Nazis tended to set up a new agency to meet each new need, an increasing number of boards, offices, commissioners, and commissioners-general have made their appearance. Except for the machinery of the statutory Reichs Agricultural Corporation, established to regulate the agricultural economy—almost a province of control in itself—the pattern of these agencies as they finally emerged is intricate, even confusing. To describe each of them would carry us into a maze of detail. The more important agencies will be dealt with later in connection with specific phases of business life and operation with which they are concerned. All we need to see here is the general pattern of overhead agencies through which it was sought to provide adequate co-ordination of the operating agencies.

*The co-ordination of controls
developed slowly.*

From the outset the Nazis "built up" both the old Reichs Ministry of Economy and the Reichsbank and linked them together. The Ministry was placed in charge of the production and flow of essential commodities. Among other duties, it supervised the allocation of foreign exchange which—under an act of December 18, 1933 —was vested in a special office and handled by bureaus attached to the regional collectors of internal revenue. It had authority over the subsidizing and licensing of exports. And through the network of Commodity Control Boards it regulated the importation and allocation of industrial raw materials and goods. The Reichsbank, on the other hand, had charge of the flow of capital and money. The two agencies continued as separate administrative entities but, except for a short interval, they had the same chief. Both were headed successively by Dr. Schacht and Dr. Funk. This personal union prevented a critical division of authority in closely related fields, and ensured effective co-ordination of basic economic policies.

However, essential policies remained outside the grip of the joint Ministry of Economy and Reichsbank; co-ordination was only partial. For example, the production, importation, distribution, and pricing of farm products were regulated by the statutory Reichs Agricultural Corporation, under the Ministry of Agriculture. Wages were controlled by the agents of the Ministry of Labor which—operating through a special Board—was also in charge of the national employment service. From November 1934 to July 1935, prices of non-agricultural domestic products (mainly prices fixed under cartel and similar agreements) were regulated by a special commissioner directly responsible to Chancellor Hitler. Likewise directly answerable to the chief of the state were the other department heads, the Inspector General for the Reichs Highways, and the Reichs Office of Space Control. In addition to the activities of these agencies, certain subsidiaries of the Nazi party frequently evolved economic policies of their own and more or less successfully interfered with the established government agencies.

In some fields, this dispersion of authority was obviously offset by agreements freely arrived at by the agencies or departments concerned, or by action taken by the Cabinet or even by Hitler himself. For example, control of prices and wages, though handled by a variety of agencies, was well co-ordinated from the outset. But in other fields there must have been a good deal of working at cross purposes. The Nazis apparently did not mind this state of affairs so long as there was economic slack to be taken up.

By the summer or fall of 1936, however, the German economic machine began to show signs of considerable strain. Unemployment had vanished in practically all the skilled trades, and a general shortage of labor was foreseen. In certain fields prices threatened to rise at an accelerated pace. Moreover, the foreign exchange problem had become more and more serious, and fear of currency inflation was spreading. To add to these difficulties, an additional expansion of production of domestic and synthetic raw materials as well as armaments was being prepared under the so-called second four-year plan. In this situation the Nazis could no longer afford to bicker or to waste government effort. It became imperative that general policies and effective co-ordination of controls be evolved.

Overall co-ordination was not achieved until 1936, when
Goering was placed in charge of the German economy.

Apparently neither the chief of state, burdened with other vital issues, nor the cabinet could do the job. It proved necessary both to concentrate and to delegate authority. On October 18, 1936 Hitler appointed Goering, his Air Minister, to control the whole economic life of the country, investing him with dictatorial legislative and executive powers for a period of four years. (On October 18, 1940 these powers were renewed for an additional four years.) Every government department or agency, the various subdivisions of the Nazi party, the groupings of chambers and of trade associations were made subject to his orders. It thus took the Nazis more than three years of steadily expanding controls to achieve what might be called an Economic High Command.

Like an army's high command, the German Economic High Command has a chief, a general staff, and a bureau. While we are well informed about the chief, almost nothing has been made known about the economic general staff, and little more about the bureau.

*The organization of the economic general staff
has been the subject of widespread speculation.*

All we know for sure is that under a decree dated December 7, 1939, Goering created an Economic General Council (Generalrat) of which he was to be chairman.[13] The Council consisted of responsible heads of the various government departments, the head of the economic services of the armed forces, a representative of the deputy leader of the Nazi party, and other persons of comparable rank. It was to hear reports of the government departments, to adjust their activities, to clarify significant problems of the war economy, and to formulate policies. Little, however, has been heard of the Council's activities.

*The Bureau of the Economic High
Command has gradually expanded.*

A little more information is available concerning the bureau of the Economic High Command. The bureau is the Office of the Four-Year Plan. Originally this office was intended to be small. Apparently it was thought that the job of formulating overall policies and co-ordinating controls did not require extensive special machinery. Goering actually started upon his new career by announcing, on October 23, 1936, that he intended to rely as far as possible on existing agencies; new agencies were not to be set up unless there was an absolute necessity.

Nevertheless, the Office of the Four-Year Plan has expanded. It would have been natural for Goering to rely on the services and facilities of the Ministry of Economy above all other agencies. But that Ministry was headed by Dr. Schacht, and Goering and Schacht did not get along well together. Dr. Schacht was too powerful a figure to be suddenly overridden, so Goering began to provide for services and facilities of his own. This trend was reversed when Schacht resigned in November 1937 from the Ministry of Economy.

There was, however, a more lasting expansion of the Office of the Four-Year Plan. In the first place, the administrative work in connection with the formulation and co-ordination of controls seems to have required more bureaucratic specialization than had been anticipated. A Deputy Secretary of State (Mr. Koerner) and the necessary staff were therefore appointed. In the second place, Goering has

[13] "Leitung der Kriegswirtschaft in einer Hand," *Der Deutsche Volkswirt,* Jan. 5, 1940, p. 424. See also Dr. Vollweiler, "Der Ausbau der staatlichen Kriegswirtschaftsverwaltung," *Reichsarbeitsblatt* No. 17, 1940, p. 277.

tended to carry his activities into the field of operating controls. For example, in the fall of 1936 price control became a crucial issue. Goering appointed a commissioner for price formation (see p. 57) and tied him in with the Office of the Four-Year Plan. Another example: in the spring of 1942 man power became a crucial issue. Again a special commissioner (see p. 38) was appointed and attached to that office.

In the beginning Goering also failed to delegate authority; he tried to do too many things himself. In grappling with technicalities, he encroached upon typically departmental duties. On November 7, 1936, for example, he issued seven decrees providing, among other things, that every entrepreneur or administration normally employing more than ten office workers should take on a suitable number of such employees above the age of forty. As might have been expected, such rambling in the field led to confusion rather than to co-ordination of authority. In 1936 and 1937 German businessmen would receive orders—sometimes conflicting—from three different government agencies, each agency claiming to be solely responsible for the handling of the matter. In time, however, Goering concentrated on laying down broad principles to guide operating procedures, rather than taking direct action himself.

Important work in formulating policies and achieving co-ordination was done by two other agencies.

Despite the high degree of concentration reached in the co-ordination of controls, the division between the operating agencies and the Economic High Command is somewhat blurred. Two operating agencies, working under the Economic High Command, have contributed vitally to the shaping of overall economic policies and to the co-ordination of controls. They are the old joint Ministry of Economy and Reichsbank and the new Ministry for Arms and Ammunition.

Prior to 1936 the Ministry of Economy and the Reichsbank, as pointed out above, had achieved a large measure of co-ordination of essential economic policies. Between them they were responsible for an enormous field—one which in the United States would roughly correspond to the activities of the Department of Commerce, the Federal Reserve Board, the War Production Board, the Office of Economic Warfare, the Defense Plant Corporation, and certain bureaus of the Treasury. In addition, the Ministry of Economy had charge of the chambers of artisans, the chambers of industry and commerce, and of the trade associations.[14] For the Economic High Command to grant a joint agency of such range a preferred share in the formulation of overall economic policies and in the co-ordination of controls was a natural development. It was delayed by the rivalry between Goering and Schacht. Later on it was enhanced by what appears to be a close relationship between Goering and Dr. Funk, Schacht's successor.

In December 1938 Goering delegated what amounted to full authority over production to Funk, who was made Economic Commissioner General.[15] This step indicated a growing significance of the joint Ministry of Economy and Reichsbank rather than constituting a clear-cut delegation of authority. The jurisdiction is not clearly divided between the Funk authorities and the Economic High Command.

For another reason—and in a different manner—the Ministry for Arms and Ammunition has also contributed to overall economic policies and to the co-ordination of controls. This agency was not established until after the beginning of the war. On February 23, 1940 Goering appointed Mr. Todt Inspector General for Special Assignments under the four-year plan. His first commission was to check continually whether measures taken by the Ministry of Economy or other government top agencies or special commissioners were being successfully carried out. Within a few weeks, however, Hitler made Todt Minister for Arms and Ammunition with a mandate to integrate the actions of all agencies dealing with production of arms and ammunition. The object was to achieve high efficiency of work and increased output.[16] To carry out his co-ordinating function, the Minister received even broader powers than those the Minister of Economy had been granted fifteen months earlier. He was authorized to issue binding orders to any agency of business or government, including certain agencies in the Army, Navy, and Air Force, and his approval was required for the appointment of top officials in agencies concerned with the production of munitions; he may even request removal of such officials. In the fall of 1942 the Ministry's powers were extended to the field. It has played a leading part in decentralizing controls under the so-called Armament Commissions that are

[14] The work of the Chambers and the trade associations in the fields both of compliance and regulation is considered at pp. 90 ff.
[15] See J. W. Hedemann, *Deutsches Wirtschaftsrecht* (1939), p. 52.
[16] Todt, who was also Commissioner General for Building Construction and Inspector General of Power Resources, reportedly was killed in a flying accident in the early part of 1942. He was succeeded in his major functions by Professor Albert Speer.

composed of regional operating officers of a number of key agencies.

As in the case of the Ministry of Economy, the breadth of these powers reveals the relative importance of the Ministry for Arms and Ammunition, but they do not indicate a definite jurisdiction or the extent of the contribution the Ministry is making to overall economic policies and to the co-ordination of controls. Goering has not abdicated as Economic High Commander, and he uses his authority whenever he desires. Whether or not in actual fact these power relationships work out smoothly is a matter of speculation. The impression is that there are loose though effective arrangements between the Funk and Speer administrations, and between either of the two and Goering.

These are the agencies that map out the policies governing control of German business. The manner in which the Nazi government acting through or under these agencies has curbed and directed the conduct of German firms will now be set forth in its major details. We propose to proceed by considering in turn each of the more important phases of business life and operation.

III. Control of Entry into Business

The Nazis have controlled entry into business. This control has been exercised in both a restrictive and an affirmative sense. The Nazis have kept out of business people and firms who wished to engage in business, and they have drawn into business people and firms who wished to keep out or who had not displayed any desire to enter. This observation applies to the expansion of existing as well as to the establishment of new firms.

Even before the Nazi regime entry into business was not wholly free. In Germany, as in other countries, a license had to be obtained before one could engage in various trades and occupations requiring special skills or high moral standards, or for the operation of plants and businesses involving dangers and risks for the community. Similarly, the German counterpart of the American certificate of convenience and necessity was needed for the opening of hotels and restaurants, bars and other drinking establishments, pharmacies, new potash mines, and a few other types of business. The purpose of these restrictions was to protect the public by keeping out of business incompetent and unreliable people, as well as inadequate plants, or to prevent ruinous competition in an extremely limited number of trades regarded as clothed with a public interest. They were not designed to effectuate broad government policies.[17] Restrictions were the exception—not the rule.

A trend toward more comprehensive restrictions began in 1932. The first important change was made when Chancellor Heinrich Bruening, acting under the legislation of March 9, 1932, prohibited the establishment of new, and the expansion of existing, five-and-ten-cent stores. In this case the German government undertook to protect a specific group of business people rather than the public. It was Bruening's intention to please the shopkeepers. He hoped to take the wind out of Hitler's best sails. The Nazis in due course far surpassed Mr. Bruening. Successively they have blocked or licensed admission to and expansion of a large number of trades and occupations.

Nazi restrictions on entry into business were designed to further various government policies.

One Nazi policy was to reward old and to build up new personal loyalties to the regime. The first step was an act of May 12, 1933,

---

[17] We are not here concerned with legal restrictions on entry into business arising from the government's monopoly of numerous public utilities, nor with factual restrictions resulting from the formation of "trusts" or other combinations in restraint of trade.

which provided that no new retail shops of any kind could be opened unless special permission had been obtained. Obviously, the favor involved for shopkeepers was much more significant than that bestowed previously. While Chancellor Bruening had intended to protect the position of shopkeepers as a class, Hitler created a safety zone around every man who happened to be keeping shop in 1933.[18] Some of the people barred from retail trade by these restrictions tried to open small mail-order houses. The Nazis' answer was an act of May 20, 1937, prohibiting the establishment of new and the expansion of existing firms in this branch of retailing. Similarly, the business of processing and wholesaling farm products was progressively subjected to entry restrictions, beginning in 1933. The licensing is handled by the proper bodies of the statutory Reichs Agricultural Corporation.

Another policy furthered by controlling entry into business was the starving out or damaging of Jews and of people politically or philosophically unreliable from the Nazi point of view. This, for example, was the purpose of compelling application for membership in the Reichs Chamber of Culture for newspaper editors and, in fact, for every person producing or distributing "cultural products."[19]

A third policy was control of investment. For example, capital should not be permitted to flow into certain domestic businesses to such an extent that they might interfere with import control. Accordingly, it was provided that after December 1935 new plants to process foreign raw materials must not be opened except by permission of the proper Commodity Control Board. It was a basic Nazi objective to preserve available capital for the rearmament and war effort. To this end, acting under the authorization of an act passed July 15, 1933, the Minister for Economy has licensed the establishment of new and the expansion of old plants in a great many industries. One condition that a proposed investment must meet is that it shall not impair government efforts to preserve the supply of man power for essential war jobs. Apparently for this purpose among others, government permission was made a prerequisite—since a decree of January 15, 1940— for the opening of new and of branches of existing wholesale trade firms (other than those trading in agricultural products and licensed by the statutory Reichs Agricultural Corporation).

[18] These restrictions were relaxed somewhat in subsequent years. They have been rigidly enforced again under an act of Mar. 16, 1939. The purpose of the 1939 legislation was to direct available manpower into industrial or agricultural occupations rather than to protect the vested interests of established shopkeepers. It is noted below, p. 82, how, finally, established shops were closed down by the thousands.

[19] As defined under the act of Sept. 22, 1933. See also p. 87.

In addition, and for a variety of reasons, the Nazis have instituted licensing in a number of other businesses. These include (1) the opening of new advertising firms, under an act of September 12, 1933; (2) the establishment of credit institutions, under the act of December 5, 1934; and (3) the opening of new, and the reopening, expansion, and conversion of old, textile plants, under an act of December 6, 1935. Moreover, a person is not free to take up or to expand farming operations when and as he likes. Under an act of January 26, 1937, the acquisition of agricultural real estate requires government permission; land acquired by a tenant under a lease agreement or for the purpose of growing and cutting timber is subject to the same restriction. In the past this permission appears to have been granted more often than necessary, thereby enabling liquid funds to escape into inflation-proof hideouts. A decree of July 28, 1942 provided that acquisition of agricultural real estate should be vigorously checked. Under regulations issued some time in the spring of 1943, a license is required even for people who, while not on the roll of a regular craft guild, are engaged in urgent maintenance work for the civilian population such as the repair of clothing and household articles of daily use or the repair of houses.[20]

The examples we have cited indicate the nature and scope of control, but the list is by no means complete. Today in Germany admission to almost every kind of trade or occupation requires some kind of government permit. The same observation applies to the operation of most types of plants, as well as to expansion of a great many existing facilities. Once the exception, licensing of business has become the general rule. Needless to add, under these conditions any right to a license has vanished. As it has been used to implement definite government policies, licensing has become solely a matter of government discretion. Whether the applicant is able to meet such requirements as may have been formulated is no longer the dominant consideration.

Existing firms have been compelled to expand their facilities and to embark upon new undertakings.

The Nazis have not stopped at *restricting* entry into business. To provide for the rapid development of the requisite war production, they have, conversely, *compelled* entry into business. Firms were forced to expand existing plant and to engage in new lines of activity.

---

[20] *Frankfurter Zeitung*, Apr. 6, 1943. It has been specified, however, that there will be no penalizing of people who start on the work without having obtained the required license.

The cases of compulsory expansion of plants are too numerous to be specified in detail.[21] Less numerous, though scarcely less important, are the cases where the Nazi government has compelled firms to engage in new types of enterprise. Two techniques have been used. (1) Certain firms or industries had to make financial commitments only. Such were the contributions toward the establishment of the Hermann Goering Works exacted in 1937 from the iron smelting and manufacturing industries. Another example was the contributions of German business toward the establishment of the Continental Oil Corporation. This company was set up on March 27, 1941 to "acquire" French and Belgian holdings in Rumanian oil companies. It was also intended that this company should operate refineries, as well as transport, process, and distribute mineral oil. While the total stock of 80 million reichsmarks was raised by the German oil industry and German banks, contributions of a non-financial character, as far as we know, have not been requested from the underwriters. As the Minister for Economy himself is chairman of the board, the government appears to have kept a strong hand in the organization of the new company.

(2) In addition to financial commitments, the other technique made firms responsible for actual establishment and operation of new business enterprise. Business was assigned new fields of activity—not merely outlets for investment. This was the favorite Nazi method of organizing the production of the principal synthetic and substitute raw materials. In this case, however, the government provided direction and advice and not infrequently subsidies as well. The first important step was taken in October 1934, when the government ordered a selected list of producers of lignite to provide the funds for, to build, and to operate plants producing gasoline and lubricants from that mineral. For this purpose a compulsory corporation was set up, the so-called *Brabag*, originally capitalized at 100 million reichsmarks. A year later the Ruhr coal mines were compelled to start producing gasoline from bituminous coal; in February 1936 action was extended to include production of oil for Diesel engines.

Similarly, the Nazis had the textile industry begin the production of staple wool, the paper industry take up production of cellulose, and the processors of vegetable oils build up a whaling fleet. Because of the absence of specific information, it is difficult to say if and to what

<hr />

[21] The significance of compulsion and the techniques used will be discussed in connection with Nazi policies to ensure reinvestment of surplus funds. See pp. 77 ff.

extent the rubber processors and manufacturers were compelled to build up the synthetic rubber industry beyond making financial contributions.

Later on, these methods were used to "develop" some of the conquered territories. For example, in the beginning of 1942, a company was established in Bremen to ensure planting and distribution of tobacco in the European East. The original capital was 180,000 reichsmarks, raised to 2,275,000 reichsmarks in January 1943,[22] which the government had the Hanseatic tobacco dealers and the German cigar manufacturers underwrite jointly. A great many such companies have been set up by German business in the course of 1942 and 1943.

These examples serve to illustrate the Nazi concept of the relationship of the citizen to his economic activities. It is not left to the citizen to decide whether or not he wants to go into or to expand his business. This vital step is a matter of government decision. For the former right to enter business as one chose the Nazis have substituted a duty to enter business as the government directs. Entry into business or its expansion can be made a matter of conscription.

Once admitted or forced into life, a business enterprise requires capital, materials, and equipment, and labor. The Nazi government has controlled the supply of all of these. It began with directing the flow of capital, of which there was very little. It then proceeded to ration or allocate raw materials and equipment, the supply of which was adequate for normal civilian purposes only. Last came labor, relatively plentiful. Though started early for selected classifications of workers, allocation of labor has been made all-inclusive since the beginning of the war.

[22] *Bremer Nachrichten*, Jan. 21, 1943.

IV. Control of Supply of Capital

Since the spring of 1933 the Nazis have directed the flow of liquid capital seeking long-term investment. This result was achieved by means scarcely perceptible to the general public. For a long time there was neither legislation nor publicity. Considerably later, it became more widely known that under a cabinet decision of May 31, 1933 a committee headed by the president of the Reichsbank had been given charge of the allocation of capital. This committee received wide powers.

Through an embargo on private securities, available
liquid capital was largely allocated to rearmament.

The chief tool of the committee was an embargo on the flotation of private securities, including those of new companies and issues to obtain additional capital. The specific purpose of the embargo was to withhold investment from the consumers' goods industries. Conversely, it has served to earmark the bulk of available liquid capital for the furtherance of the rearmament and war effort. This was done in two ways.

The first way led through the Reich Treasury. As the embargo blocked the flow of capital into private issues, a more or less monopolistic market for government bonds was automatically created. It might seem that there was no necessity for such a market, since the purchase of government bonds was made mandatory. While it is true that the government "placed" billions of loans "directly" with insurance funds, savings banks, and other reservoirs of capital, originally it also offered large amounts for public subscription. Actually, the embargo on competing private issues has been an important factor in the underwriting of government issues, which has been very satisfactory indeed.[23] Needless to add, the funds thus raised by the Treasury soon found their way into the numerous channels associated with the war effort.

The second way avoided the treasury circuit and was the Nazi equivalent of a more normal method. The embargo on private issues, it should be noted, is not an absolute one; it may be lifted. Not that an applicant may claim that it be lifted on the ground that

[23] For the period from 1926 to 1932 public securities issued totalled 3,902 million reichsmarks; the corresponding figure for the period from 1933 to April 1939 was 16,346 millions. The obligations issued during the latter period were mainly 4½ per cent Reich loans and Reich treasury bonds. These figures and those in the following paragraphs were taken from or based on figures in Reichskreditgesellschaft, *Economic Conditions in Germany in the Middle of the Year 1939*, pp. 52, 53.

he had complied with definite legal requirements. The Nazi government is free to grant permission to issue securities to one business enterprise and to refuse it to another. It upholds or lifts the embargo at its discretion. The German embargo should not be confused with control of the issuance of securities by the Securities and Exchange Commission in this country. The Nazis are not concerned with protecting the investor. They are concerned only with the investment. They direct available supplies of liquid capital to the proper channels.

In practice the Nazis have lifted the embargo for a number of firms in a few essential industries only. These include chiefly: ore and coal mining; iron smelting and manufacturing; production of synthetic and substitute raw materials, especially of gasoline; and the power and chemical industries.

Private issues showed an upward trend after 1933, but they were greatly surpassed by governmental issues.

Originally, private stock and bond issues authorized were few and far between. Annual totals were: 93 millions reichsmarks in 1933; 147 millions in 1934; and 159 millions in 1935. In order to prompt industrial expansion under the so-called second four-year plan, the Nazis lifted the embargo more frequently in the following years. Total yearly issues rose to 472 millions in 1936; to 591 millions in 1937; and 689 millions in 1938.[24] For 1939 they dropped again to 511 millions,[25] thereby dashing the hopes held out by the Minister for Economy in advertising his "New Financial Plan." Total private issues for the seven years 1933-39 were only about 2,662 million reichsmarks, excluding the Reich's subscription of 240 million shares in Reichs Works Hermann Goering. This was not much more than one-third of the total of 7,336 millions for the seven years 1926-32 preceding the Hitler regime.

The extent to which the Nazi government has controlled the allocation of capital is even more apparent when one compares the figures for private issues with public issues. For the seven years before Hitler new private issues accounted for 65.4 per cent of total long-term issues; for the period from 1933 to April 1939 private issues represented only 13.1 per cent of the total.

[24] This figure does not include 240 millions of 395 millions issued for Reichs Works Hermann Goering which was subscribed by the government.
[25] E. Noelting, "Kredit, Finanz und Steuerpolitik nach Kriegsausbruch," *Der Deutsche Volkswirt*, Dec. 15, 1939, pp. 307, 309.

The war years saw a considerable rise in the volume of private issues. Total issues were 1,759 millions in 1940; 2,418 millions in 1941; and 1,868 millions in 1942.[26] However, the percentage of private to public issues continued to decline in consequence of the stupendous war requirements.

*Using a variety of methods, complete control
of investment capital was obtained.*

Government limitation and allocation of the issues of securities did not constitute complete control of the flow of investment capital. To approach this goal the Nazis have also restricted investment in mortgages.[27] In addition, the various techniques of licensing both new and expansion of existing facilities, as well as government allocation of raw materials, equipment, and labor, have indirectly checked normal capital flows. Their net effect has been to leave only such opportunities for investment as the government deemed proper. As early as the summer of 1935, about two-thirds of all investments were reported to be either directly or indirectly under the "influence" of the government.[28] If this was the situation in the initial stage of the militarization effort, it may safely be assumed that government control of the flow of capital has long since become all-inclusive.

[26] *Deutsche Bergwerkszeitung*, Dec. 10, 1942. The figures for 1940 and 1941 probably include some refunding issues. The largest part, however, represents new plant, new equipment, improved methods of production, and so forth (the government takes care of damage due to enemy action). Together with other indications, they suggest that, contrary to what is widely assumed in this country, the German industrial war potential was still expanding at the end of last year.

[27] Information as to the exact techniques employed is not available to the writer.

[28] *Vierteljahrshefte Zur Konjunkturforschung*, 1935-36, Pt. A, p. 181. See also Kenyon E. Poole, *German Financial Policies 1932-1939*, note 32, p. 218.

V. Control of Supply of Materials

Prior to the Nazi regime, the flow of materials was subject to certain import restrictions resulting from foreign exchange rationing. These restrictions, introduced by Chancellor Heinrich Bruening in 1931, were insignificant, especially when compared with what was to follow. Importers were allotted foreign exchange—figured on a percentage of their former imports. There were no restrictions on the goods that could be imported nor on their origin, other than those imposed by tariffs and national quotas.

Control of materials developed gradually both as to choice of goods and machinery of allocation.

For the free flow of materials the Nazis have substituted allocation by the government. The objective was to meet the direct and indirect requirements of the rearmament and war effort. Imports were first subjected to control. In 1933 imports of dairy products, oils, fats, grains, feed, and eggs were regulated, but import control was made all-inclusive by Schacht's "New Plan" announced in the fall of 1934. In order to restrict imports of goods not essential for the rearmament and war effort, foreign exchange was made available for authorized imports only. Moreover, a decree of June 24, 1935 prohibited unauthorized imports.

Restrictions on imports increased the demand for domestic goods, threatened to produce bottlenecks, and caused businessmen to hoard scarce materials. As a result, domestically produced goods had to be included within the ambit of control. Decrees and regulations, issued under the basic act of March 22, 1934 and subsequent legislation of September 4, 1934, covered a number of essential raw materials only. As additional shortages developed, control was progressively extended to practically all raw materials and finished products.

Elaborate machinery was established for the purpose of handling allocation. Between 1933 and 1936 five Commodity Control Boards (Reichsstellen) were set up to deal with specified classifications of farm and truck garden products. Beginning on March 26, 1934 and ending at the outbreak of the war, 26 additional boards (Ueberwachungsstellen, also named Reichsstellen since August 1939) were set up. Each of these boards was assigned specified industrial, mineral, chemical, or other materials. The network of 31 boards thus established covered every commodity for which there are statistical records. The five agricultural boards concentrated on imports, while control of domestic agricultural products was left for the most part to the marketing associations of the statutory Reichs Agricultural Corporation. On the other hand, the non-agricultural boards controlled the flow of both imports and domestic products. To some extent the boards acted through the statutory trade associations. They also acted through existing cartels, which found a new field of action in allocating materials to their members.[29] Eventually the jurisdiction of some of the boards was overshadowed by other agencies.

A variety of methods have been used
in allocating materials.

The experienced heads of the various boards operating under the Ministry of Economy were permitted considerable leeway in working out methods that were administratively suitable to the several commodities and industries, and to change their methods as need arose. The resulting pattern is highly diversified and very fluid. Our knowledge of the methods followed is incomplete. Some boards have published hundreds of regulations; others have published none; no annual reports have been made available to the public. Leonhard Miksch, an outstanding and penetrating observer enjoying all the advantages of being on the spot, regretfully admitted as early as 1937 that hardly anyone had a complete picture of the work of the boards.[30] If anything, the view has become further obstructed in recent years.

Prior to actually allocating materials, the Commodity Control Boards made statistical surveys of existing inventories. Detailed questionnaires assembled information covering types or kinds of a commodity in stock, dates of purchase, existing agreements to buy and sell, and other pertinent data. Next, control of inventories was undertaken to prevent either over-stocking or under-stocking. Regulations were issued to ensure adequate inventory bookkeeping, especially for goods easy to hoard. *Maximum* stocks that might be held were set in most cases, but as early as 1935 iron and steel producers—and subsequently other producers requiring essential imported materials—

[29] As a result of government control of prices, the cartels have been increasingly curtailed in their traditional role as agents of market control. See below pp. 49-50.
[30] The following survey is based largely on a number of articles by Leonhard Miksch. "Praxis und Wirkung der Einfuhrueberwachung," *Wirtschaftskurve*, I (1937); "Die Wirtshaftskontrolle der Ueberwachungsstellen," the same, II (1937); "Die Auftragslenkung," the same, III (1939); "Bewirtschaftungskartelle," the same, I (1940); "Rationierungssysteme," the same, III (1940); "Das 'System Speer'," the same, II (1942); "Von der Reichsstelle zum Lenkungsbereich: Zur Reform der Kriegswirtschaft I," the same, IV (1942).

were forced to provide for specified *minimum* inventories, probably so as to be able to meet any emergency in connection with the armament program. To facilitate enforcement of these policies, firms are requested to submit periodical reports covering their stocks and other details.

Although our information is not altogether adequate, three successive but overlapping phases in the development of methods of allocation can be distinguished.

In the first phase, reliance was placed primarily
on a rather flexible system of rationing.

Under the original scheme—used chiefly for raw materials—consumption in some base or reference period was used as a yardstick. The boards restricted the use of scarce raw materials to stated percentages of base-period consumption. They then regulated either the buying or the processing of controlled materials, or both. As early as 1934 the buying of wool, cotton, hemp, flax, jute, asbestos, and scrap paper was regulated. For some commodities each purchase required an authorization; for others blanket licenses were issued permitting purchases to the full amount of the authorization within a specified period. The processing of materials was controlled in the textile and metal industries, as well as in the case of rubber, cigar tobacco, sulphuric acid, and other goods. Control of processing facilitated the inclusion of inventories in the data used in determining allowances; it became also an important instrument in effecting the gradual conversion of industry to the use of domestic substitutes in place of imported materials.

Rationing in terms of uniform specified percentages was never applied. Neither was rationing used as an exclusive method of control. In practice, rationing was modified and supplemented by other methods of allocation. In order to provide flexibility, the base or reference period was not uniform throughout an entire industry. Moreover, in the numerous cases in which new plants were opened or existing plants expanded, the reference period had to be fixed more or less arbitrarily. Thus, essential industrial operations received relatively more raw materials than others. In addition, producers of certain articles received extra or special quotas over and above allowances under rationing. For instance, this was done to increase exports, which in turn made possible additional imports of essential materials. It was also done, on a large scale, in order to promote production of arms and ammunition. Broadly viewed, rationing of raw materials was intended to and did change the composition of the output of finished products rather than restrict total output. Raw materials were released for the production of goods deemed essential for the rearmament and war effort.

With the same objective in mind, the Nazis have limited or prohibited certain civilian or non-essential uses of scarce materials. The first developments along this line date from the middle of 1934, when restrictions were placed on the use of lead, copper, tin, chromium, mercury, and cobalt in the fabrication of a variety of products. As time passed, restrictions were imposed on the use of more and more materials in an increasing number of products. The boards have used two techniques. In the case of non-precious metals they originally prohibited specified uses only, other uses remaining lawful. Since October 1, 1939, however, there has been a blanket prohibition on the use of aluminum, magnesium, and other metals, which can be used only after special permission has been obtained.[31]

Scarce materials have also been saved by increasingly restricting the output of specified goods. Besides prohibiting many types of goods altogether, others must not be made unless production follows authorized standards. Thus, specific chemical or mechanical methods have been prescribed for the processing of certain materials. Similarly, standards have been made applicable to final—both capital and consumers'—goods. In the capital goods field a striking example is the progressive reduction in the number of types of machines. A total of 3,637 types of machines of all descriptions, built at unspecified dates, had been reduced to 1,011 by September 1942. Another example is the electrical industry, where the number of types of insulators for high voltage outdoor lines reportedly had been reduced from 120 to 16, and 4 standard designs for low voltage branch boxes had been substituted for the 500 hitherto in use.[32] In the consumers' goods field only one style is authorized for stiff collars and nightshirts for men and three styles for baby outfits (regulations of June 1942); only 15 types of pencils, copying pencils, color pencils, and drawing pencils, most of them to be unvarnished (regulations issued on December 31, 1942 by the committee on army and general equipment); and only one type of perambulator instead of 132 models made formerly.[33]

[31] *Der Deutsche Volkswirt*, Sept. 29, 1939, p. 2467.
[32] *Deutsche Bergwerkszeitung*, Sept. 24, 1942, Dec. 2, 1942.
[33] *Berliner Boersen Zeitung*, Mar. 18, 1943. In addition to saving materials, the progressive

Within the framework of percentage rationing, quotas, limitation orders, and standards, a firm was free to choose its product and its customer. Originally this situation apparently did not raise serious problems. On the whole, there was leeway enough for government orders and civilian contracts to be filled adequately. As rearmament was stepped up, however, and full capacity reached or approached, orders appear to have piled up. Procurement agencies began to bid against one another, in addition to competing with civilian needs. At this juncture it became necessary for the government to adopt a first-things-first policy; that is, to differentiate between contracts on the basis of urgency and national significance.

A system of priorities was the distinguishing feature of the second phase.

Establishment of an order of preference for the execution of contracts characterized the second phase in the process of allocation of materials. Exports were given a particularly high ranking. Whenever possible, they were granted priority over other essential commitments. While every export transaction required specific permission, techniques of securing priority changed. Originally, quotas (at least extra quotas) were allocated to exports as orders from abroad were received. This method, of course, slowed up the carrying out of such orders. Submission of the required evidence to the proper Board and its review in order to prove the existence and to verify the particulars of an export engagement frequently took a considerable length of time, even though the goods ordered were urgently desired. It would seem that firms known to be trustworthy in their use of scarce materials were permitted advanced production for an export inventory, so as to be able to fill orders without loss of time. Moreover, toward the end of 1938—when stricter controls of the foreign exchange proceeds and increased profits in domestic trade lessened industry's interest in exports—it was made mandatory for a number of industries to earmark a specified fraction of total capacity for exports.

---

standardization of techniques of production and of products saved man power, fuel, power, and tools. Indeed, it should be seen as part of the drive, strongly spurred by the Nazis, to "rationalize" practically every technique of production and administration down to the wording of business terms and contracts, and extending to every branch of industry, trade, banking, insurance, and farming. Contrary to a view widely held in this country, the process of rationalization continues unabated in Germany. Almost every issue of the more important daily papers and magazines read by business people and engineers contains items showing that a continuous effort is being made to reduce further the number of sizes and styles of goods produced, and to achieve a more scientific management in practically every line of business enterprise.

In the domestic field, the Boards have issued preference orders for selected materials, grading contracts in two different ways.

(1) Originally they granted a priority when a particular contract seemed to call for it. The proper Board issued the license after satisfying itself that the demand was warranted. But not all licensed demands have been treated on an equal footing. Two types of license—ordinary and urgent—have been issued. The supplier is required to give preference to the latter. Another and more informal technique involved direct interference by certain Boards with delivery schedules. This apparently was done to some extent in the case of metals, and also for staple fiber, paper, cardboard, and other materials.

(2) The second method was the advance establishment of regular preference ratings through code numbers (Kennziffern). It was first introduced on May 1, 1937 to guide manufacturers of iron as to which orders to fill first. Apparently it was also used for other metals and timber; and in 1939 it was extended to textiles. Under this scheme the original method of rationing materials for, or of allocating (advance) quotas to, the manufacturer was dropped, except when necessary to provide adequate maintenance and repairs as well as to meet recognized civilian needs. For government contracts involving the manufacture of iron, the government rather than the manufacturer was rationed. The available or expected supply of iron was divided among the armed forces, the Nazi party, the Labor Service, the Minister of Transportation (for the Reichs railroads), the Commissioner for Building Construction, and others. Each claimant received an allotment of iron for a specified period. From its allotment a given agency, such as the armed forces, released to a manufacturer the amount of iron required for filling an order. The contractor, in turn, manufactured or at least delivered the goods in the order indicated by the code rating attached to the contract.

This scheme of grading contracts requiring consumption of iron appears to have served its purpose. It was reported, however, that the leeway was such that the claimant agencies could exceed their allotments. For this and perhaps other reasons deliveries frequently got off schedule. After experimenting at some length with other devices, the Board for Iron and Steel, acting under regulations issued on June 13, 1942, set up a central clearing office for the distribution and allocation of iron (Eisenverrechnungsstelle). This office opens an "account" for each claimant agency. On the "credit" side of each account there is entered the amount of iron allotted. Prior to placing a contract involving consumption of iron, the agency draws a "check" against its account. That check cannot be "cashed"—which means the iron must not be manufactured—unless and until the clearing office certifies that the agency had a "balance" sufficient to cover the quantity of iron the contract requires. This device, it was hoped, would forestall consumption of iron in excess of the total amount released. At the same time, manufacturers were enjoined from accepting more orders than they could reasonably expect to fill within prescribed delivery schedules.

We are not in a position to appraise the effectiveness of this improved scheme. Whatever its merits, it was obviously not intended to guarantee that all raw or processed materials entering the fabrication of a given iron product be available to the manufacturer or available when they were needed. This, however, was precisely the problem of co-ordination that in the meantime had come to the fore. Conceivably, the Nazis might have attempted to solve it by extending the technique of preference ratings from a few selected materials to cover contracts for all or the majority of scarce materials, and by refining it at the same time. Instead, they decided that it was the duty of the government—or some authorized agency under the government—to figure out the amount of each material required for the production of certain goods, and to allocate the amounts directly.

Unified allotment of all materials required for a given contract or operation marked the third phase.

Attempts to carry out this policy are the distinguishing feature of the third phase in the allocation of materials. In view of the great variety of materials required for each project, it is not surprising that the first attempt at this inclusive type of control was made in building construction. Sometime in 1939 it was ruled that to be able to get the required materials a contractor had to apply to one of the 19 agencies set up for the purpose. For each authorized building project, the agency was to allocate at one time the required amount of each material involved—timber, iron, concrete, and the like.

Another attempt was the "System Speer"[34] initiated in the spring and summer of 1942 in order to smooth the production of war materials. Under this scheme, two sets of organization were established, one for the end product branches, the other for the raw material branches concerned. For each important end product, ammunition, warships, tanks, etc., a committee of experienced specialists (Hauptaus-

[34] Named for Professor Albert Speer, Minister for Arms and Ammunition.

34

schuss) was appointed together with special committees and subcommittees (Unterausschuesse, Arbeitsausschuesse) to deal with the various sub-types involved. The end product committees are matched by similar committees (Hauptringe, Ringe) in the production of raw materials and accessories. Both networks of committees are topped by the Reichs Minister for Arms and Ammunition. The end product committees have full control over production programs of firms under their jurisdiction, including the right to specify which firm is to specialize on a given final good, and even to prescribe the manufacturing processes applied. They have also made an important contribution to the progressive reduction of types and sizes.

Thus endowed with broad powers, it appears they draw up production plans that will be submitted to the material branches committees, and modified and correlated in the light of data gathered on raw materials and facilities available. Eventually, they distribute the final government orders to the various firms and claim the necessary amounts of raw materials—directly, so it would seem—from producers.

Similar attempts have been made in other lines. A striking example is machine building, where the special commissioner—as he stated[35]—periodically lays out production programs for each factory under his jurisdiction, to which required materials are allocated. At least this is true for large items such as steam shovels, dredging machines, and elevators used in mines. More "fields of control" (Lenkungsbereiche) have been established since the fall of 1942.

Clearly, under these schemes the Commodity Control Boards, organized along the lines of raw materials rather than of end products, have lost a great deal of power. Some have been abolished. Others have been merged, an achievement that must also be seen as part of the general drive to reduce the number of offices and to effect a saving in man power. Still others have been replaced by so-called Reichs Associations (Reichsvereinigungen) set up progressively since April 1942 in an apparent attempt of the government to enlist more effective co-operation of business in organizing production of and in allocating coal, iron, and other materials and certain finished products.[36]

[35] Karl Lange, "Kriegswichtige Maschinen," *Deutsche Bergwerkszeitung*, Aug. 16, 1942.
[36] The relationship of the Reichs Associations to the statutory trade associations, to the remaining cartels, to the Speer Committees, and to the various special commissioners is anything but clear. There are good reasons to suspect that some of the Reichs Associations owe their life to the personal ambitions of certain high officials in the Ministry of Economy rather than to actual economic necessities.

This third phase of government allocation of materials brings our analysis up to date. It should be noted, however, that the devices summed up under the second and third phases regulated principally the allocation of materials used in making capital goods and goods required for the conduct of the war.

*Special methods, including ration banking, were used
to control the distribution of consumers' goods.*

The *production* of consumers' goods has been increasingly restricted since 1934, through rationing of materials for and allocating of quotas to producers, as well as by the ever-growing number of limitation and standardization orders. But *distribution* of consumer goods through the channels of wholesale and retail trade, by and large, was free until the beginning of the war. As the Nazis have, since that time, progressively cut down civilian consumption to the most essential requirements, it became inevitable that distribution of consumers' goods to wholesalers and retailers be regulated in turn. For this purpose definite methods of allocation were evolved. To round out the picture, these will be considered briefly.

Originally retailers received more or less rigid quotas. The drawback of this method was that it did not take into account changes in turnover. As the retailer won new customers or lost old ones, his designated quota turned out to be either too small or too large. The problem clearly was to adjust allocations to wholesalers and retailers in accordance with authorized consumer demand. Two methods of fitting allocations to the yardstick of consumer demand have been used.

The first and more common involved the successive use of three devices: ration cards or purchasing licenses for the consumer; retailers' purchasing licenses; and wholesalers' purchasing licenses. To the extent that goods were represented by coupons or consumers' purchasing licenses, the retailer was given a license to buy from the wholesaler, who in turn was licensed as to his own purchases.

The other method of allocation has been applied since late 1939 or early 1940 in order to facilitate the smooth distribution of wearing apparel. A single device—ration points—is used to supply consumers, retailers, and wholesalers. Consumers receive clothing ration cards of a certain number of units, which are surrendered to retailers according to point ratings specified from time to time—so many points for a shirt, so many for a coat, and so forth. The retailer deposits the units he receives to his "point account" with a clearing agency set up for the purpose. To be able to buy from a wholesaler, a retailer draws a "check" which is cleared only if the necessary balance is available. The identical procedure is followed in allocating goods to the wholesaler,[37] but his credits are the retailer's checks rather than units from consumers' ration cards.

In one way or another the Nazis control the supply of materials to every firm, while the flow of a number of materials is regulated in its entire course from the producer or importer of the raw material to the consumer of the finished product, either a government procurement agency or the individual consumer. To be sure, the goods are not invariably delivered at the scheduled place and time, nor do the quality and quantity always conform with the orders and the buyer's expectations. As a result, stoppages or lags in production doubtless have occurred, especially under the strain of recent years. By and large, however, government regulation of the supply of materials to business enterprise has been effective. In part, this was due to control of prices, which removed an opportunity for the bidder to switch the goods from the channels planned or authorized by the government. Both the allocating and the pricing of goods, while carried out in the main by different agencies, were closely knit in one commodity policy. Control of the supply of materials to business enterprises has certainly facilitated control of prices, but the converse is also true; price control has checked the unregulated flow of goods, thereby helping to control the supply of materials. It can thus be seen that the suggestion, often repeated, that in the Nazi economy government distribution of commodities has *supplanted* the function of price is an over-simplification. Price has been *used* by the government to promote a desired allocation of commodities.

[37] This method later on served as a model for the allocation of iron under the regulations of June 13, 1942 discussed above.

VI. Control of Supply of Labor

Before the Nazis came into power, the German worker was free to accept or reject employment and to relinquish his position whenever he liked. Under the Constitution of 1919 and an act of 1920 there were minor restrictions for the employer as to his right to discharge.

For the free flow of labor the Nazis have substituted allocation by the government. The principal administrative agency is the statutory Employment Service Board established in 1927, which operates through a nation-wide network of 491 (July 1942) regional and local offices. An act of November 5, 1935 gave the Board a monopoly of employment service, vocational guidance, and the placing of apprentices. Later the Board was incorporated in the Ministry of Labor, but on March 21, 1942 a personal decree by Hitler placed it under the orders of district leader (Gauleiter) Sauckel, who is known for his ruthless efficiency. Sauckel was commissioned to centralize the mobilization and allocation of man power on a continent-wide scale.

As in the case of the allocation of materials, allocation of labor was a gradual development both with respect to classification of workers and rigidity of methods. In contrast to the allocation of materials, however, there is a single pattern of control of labor and, as it finally evolved, it is perfectly transparent. The Nazis have allocated labor in both restrictive and affirmative ways. They have tied workers to their jobs; they have moved workers from one job to the other.

The establishment of a vocational record for each worker greatly facilitated control of labor.

Between June 1935 and September 1936 the employment authorities introduced the Work or Employment Book for every wage–or salary–earning German. In the spring of 1939, this requirement was extended to workers employed on the basis of a family relationship, as well as the independent businessman. At a later date somewhat similar documents were issued for war prisoners and foreign civilian workers. The employment book is a full vocational record—always kept up to date—of the person concerned. A transcript of every book is on file with the proper employment office. Thus the employment authorities are in a position to make at any given time a complete survey of the skills and abilities available in Germany, as well as to ensure that every man is working in a position for which they consider him well fitted. Undoubtedly these employment books have also facilitated the work of the draft agencies in selecting and deferring essential workers.

*Labor was at first subjected only
to control of hiring.*

As to tying workers to their jobs, the Nazis originally confined themselves to controlling the hiring of workers who wished to move. Although there was a reservoir of unemployed estimated at 6 to 7 millions when Hitler seized power, the great construction projects incident to rearmament attracted so many unskilled workers from the country that farms began to run short of hands as early as 1934. As a consequence, a law was enacted in May 1934 providing that a person engaged in agriculture at a specified date or within a specified period would not be permitted to accept employment on a non-agricultural job except by special permission of the proper employment office. Meanwhile the manufacture of equipment and of actual war materials had begun. By the end of 1934 it had progressed to a point where scarcity of metal workers was noticeable, especially in the case of those classed as skilled. As a result, skilled metal workers began to move throughout the country in search of better jobs, a condition that interfered with productivity in certain districts. A decree was therefore issued on December 29, 1934, which made it illegal for any public as well as private business or administration to hire skilled metal workers residing outside the district of their original employment office unless permission had first been obtained.

As the manufacture of arms and ammunition was stepped up, "pirating" of metal workers became frequent. To meet this situation, employers in certain industries entered into agreements in 1936 not to hire a worker unless he could show a certificate of release from employment issued by his former employer. When this method proved inadequate, it was decided early in 1937 that henceforth permission must be obtained from the proper employment office by any public or private business or administration desiring to hire a metal worker, skilled or unskilled. In October 1937 hiring of masons and carpenters was similarly restricted, while the following May the influx of any kind of labor to the building trades with their relatively high wages was made subject to control by the employment authorities. When full employment was reached, control of hiring was extended to cover many additional classifications of workers. On March 10, 1939 the Nazis decreed that enterprises of any kind (including households) were not to hire workers or clerks engaged in agriculture, forestry, mining, the chemical industry, and the production of building materials, except with permission of the employment authorities.

Clearly, control of hiring was a considerable check on the mobility of labor. But workers were not tied firmly to their jobs. Under the legislation of May 1934, a farm hand was free to accept employment on another farm. Under the decree of December 29, 1934, a metal worker was free to change positions within the district of his employment office. Under the decrees of February 11 and October 6, 1937, metal workers, masons, and carpenters were free to take jobs, for example, as clerks or seamen. Nor did control of hiring prevent any of the types of workers we have mentioned from giving up working for wages, from becoming independent businessmen, or from retiring.

In due course control of release was added to control of hiring.

To meet this situation the decree of March 10, 1939 added control of release to control of hiring. After this date release of workers employed in specified trades was prohibited unless permission of the proper employment office was first obtained. Henceforth a change of employment was possible only after two employment offices approved: (1) the office in charge of the employer whom the worker wanted to leave; and (2) the office in charge of the employer for whom the worker wished to work. Workers in the classifications coming under the decree of March 10, 1939 were thus fully tied to their jobs. Only the employment authorities could release them.

The classes of workers affected of course did not represent the whole of the German labor force. The remaining classifications were tied to their jobs at the beginning of the war. A decree of September 1, 1939 extended the prohibition against hiring and releasing (except by permission of the employment office) to all enterprises, businesses, administrations, and so on, private as well as public, and also to all workers, clerks, apprentices, and persons working without claiming wages or salary.

In this network of ties only one weak spot remained. The labor contract could be dissolved by *mutual* agreement of employer and employee. Originally it was thought that this freedom was merely a technical one, inasmuch as few employers could be assumed voluntarily to release such workers as they were lucky enough to have, and as any employee actually released had to report forthwith to an employment office. It would seem, however, that rather frequently workers who were dissatisfied with their jobs deliberately conducted themselves in such a way as to make the employer wish to get rid of them. Furthermore, a good many discharges of workers must have been required by reason of stoppages caused by lack of raw materials or plant destruction by aerial bombardment. Apparently quite a few employees thus released by mutual agreement managed to slip through the mesh of the employment authorities. In order to check this trend, Mr. Sauckel ruled on May 20, 1942 that in specified war industries dissolution by mutual agreement of the labor contract of permanently employed adult men was unlawful and that only the employment authorities could cancel the contract. At the same time he ordered employers to report to the proper employment office all workers that could be spared. On September 29, 1942 similar regulations were issued applying to women, juveniles, and temporary employees.

It appears that on the whole the employment agencies have been successful in tying workers to their jobs. It should be noted, however, that to a large extent this result was due to the fixing of standard and maximum wages by the labor trustees. We have seen that allocating and pricing of goods were knitted in one commodity policy. Likewise, regulation of both employment and wages, while handled by different agencies, was closely integrated in one labor policy. Complete control of hiring and release has certainly contributed to keeping wages stable, but the converse is also true. Stability of wages has certainly acted as a check on the mobility of labor, thus helping to achieve complete control of hiring and release.

Compulsory mobility has replaced the
mobility of a free labor market.

At no time was it the intention to tie workers to their jobs to a point where the whole German employment structure would be frozen. Rather, it was a prerequisite for successful redistribution of labor—by or in behalf of the government. Indeed, far from doing away with the mobility of labor, the Nazis have provided labor with a new type of mobility. For the classical mobility of labor as expressed by the worker's right to quit and to seek employment elsewhere, and the employer's right to hire and to fire, they have substituted compulsory mobility.

The object of compulsory mobility was to make certain that workers were used to their fullest productivity. There were quite a number of workers who for one reason or another had left their original trades and taken jobs in other lines or wherever they could find employment.

Such changes had been common in the years of the great depression. The Nazis took the view that the worker was most productive in the field in which he had been trained. Accordingly, they decided to reallocate labor on this basis, on the whole regardless of whether or not a worker had become adjusted to his current occupation or to his surroundings.

Originally the Nazis reallocated selected groups of workers only. Thus, under a decree of November 7, 1936, former metal and skilled construction workers were urged—and many of them forced—to leave their current jobs and to accept jobs more nearly in conformity with their former occupations. Similarly, toward the end of 1938 and early in 1939 all former miners engaged in occupations other than agriculture were ordered to return to the pits. By April 1939 the transfer of workers to their original or previous occupations ceased to be merely a matter of expediency designed to meet urgent needs in a few industries. It became a principle of labor allocation applicable to every worker. Nor was this all. To promote productivity, workers have even been moved within their regular or finally allocated occupations. Thus, progressively since the middle of 1938, the employment offices have endeavored to make sure that skilled men were put on skilled jobs only. If an employer does not have an adequate job to offer, the skilled man is transferred to an adequate job with another employer. These policies have also helped to prevent hoarding of workers by employers.

Much larger numbers of workers have been moved around to fill gaps in the supply of labor when and as they occur. Again the Nazis began by moving only selected groups or classes of workers. For example, in 1937 employers usually employing more than ten clerks were forced to take on a suitable number of clerks above the age of 40. The compulsory employment of older clerks was intended to, and actually did, lead to the discharge of roughly the same number of younger clerks who became available for manual work on farms, on construction projects, and in other critical occupations. Similarly, in 1937 textile workers below the age of 30 working on part time were deprived of relief, thereby putting pressure on them to enter other occupations. It should be added that in the following years hundreds of thousands of peddlers, itinerant salesmen, shopkeepers, and artisans, were assigned to jobs as workers in critical industrial occupations and on the farms. This process was considerably speeded up under the strain of the Russian campaigns. Following a decree of January 27, it reached a climax in the spring of 1943 when it proved necessary to provide for wholesale replacement of workers of military age and fitness that were drafted into the armed forces.

As labor scarcity became more general, shifting of workers on the basis of the group they originally belonged to was supplemented by transferring individual workers whenever they happened to be available. Under an act of June 22, 1938, as amended February 13 and March 2, 1939, every German and a great many aliens could be conscripted to work for an indefinite period on any job to which they were assigned by the employment authorities. Labor conscription has been used to mobilize people who were not gainfully employed, including among others hundreds of thousands of women (an estimated 1 to 2 million in the spring of 1943 alone). It proved equally useful as a technique for transferring workers from whatever jobs they had to other jobs where they were urgently needed. Thus, tens of thousands of workers were called up in the summer of 1938 to work on what was to become the Siegfried Line. Later on, in a circular dated July 9, 1939 the Minister of Labor went so far as to suggest that conscription be used to send back to the farms wives of farm workers who had preferred jobs in trade and industry. Since the war conscription has gradually developed into a kind of panacea, available at the convenience of employment authorities whenever a gap of some importance was to be filled. Specifically, since 1940-41 wholesale action has been undertaken both by the employment offices and by special mixed commissions to strip all plants of non-essential workers and to assign them to jobs on war production.[38]

In Nazi Germany the flow of labor is completely under government control; tying the worker to his job and moving him around are merely two phases of one process. It can thus be seen that under the Nazi system control of labor is essentially similar to control of capital and of materials; the difference is in the techniques used rather than in the completeness of domination achieved. The whole labor force, the whole reservoir of goods, and the total of capital available in and for Germany should be considered as three big pools from which at its discretion the government assigns quotas to businesses and to which such quotas may be returned when the government dictates.

The apparent success of these policies, as has been suggested, was largely due to control of various financial aspects of business operations involved. These aspects will now be discussed.

[38] For a more comprehensive analysis of policies in allocating labor, see L. Hamburger, *How Nazi Germany Has Mobilized and Controlled Labor* (The Brookings Institution, 1940).

VII. Control of Prices

Price regulation by both governmental and non-governmental agencies was fairly widespread before the Hitler regime. Wages were set predominantly by collective agreements freely concluded between organizations of employers and organizations of workers. When agreement was not reached the government frequently arbitrated. The government also had the authority to extend wage agreements and arbitration awards to cover entire trades and industries. Prices of commodities and services were regulated to a large extent by cartel agreements.[39] Membership in two cartels—coal and potash—was compulsory, and the government controlled their price policies. In addition to cartel agreements, there were numerous retail price maintenance agreements. Prices in the numerous government-owned industries or enterprises were, of course, set by the government or under its authority. In the public utility field, prices were controlled by the federal states and by the municipalities.

Thus, groundwork had been laid for more extensive price control. It was of great value in 1931-32 to Chancellor Heinrich Bruening who, in pursuit of his deflationary policies, reduced wages, interest rates, rents, and the prices of a great many commodities. It was of equally great value as a basis for Nazi price control.

CONTROL OF BASIC COST ELEMENTS

From 1933 to late in 1936 Nazi price control centered on wages, farm products, a great many raw materials and semi-finished goods, and imported commodities.

Control of wage rates and wage incomes
were the first steps in price control.

The Nazis started price control by controlling wages. To control wage rates they used wage determinations (Tarifordnungen) to cover entire trades and industries within a given district or throughout the whole Reich. From 1934 on, these determinations, as established by Labor Trustees who acted under the Ministry of Labor, replaced freely concluded collective agreements. In the early stages of recovery, with mass unemployment still prevalent, wage determinations were intended to be and actually were minimum rate regulations. But

---

[39] In 1930 Professor Ernst Wagemann estimated that about 50 per cent of German basic material production was cartelized. *Struktur und Rhythmus der Weltwirtschaft*, p. 277.

in time this changed. With the militarization effort the demand for labor rose sharply and workers might have obtained higher wage rates, if they had been free to act. The Labor Trustees, however, refused to raise the rates set in the original schedules; exceptions were made only for a limited number of workers' classifications. In this manner minimum rates gradually became standard or normal rates, stabilized for the most part at the 1933 level.

The stabilization thus effected was not altogether rigid. Individual employers were free to offer rates above standard, and employees were free to accept them. When by the beginning of 1938, such higher rates tended to be widely paid despite official discouragement, the Labor Trustees were authorized to declare them to be maximum rates (June 1938).

For a while employers found ways and means of increasing the compensation of individuals (over and above increases due to lengthened hours); while abiding by the scheduled maximum rates, they granted a variety of bonuses. During the year 1939, however, the government tightened up the loopholes as they developed and since the first months of the war the maxima have been enforced with the utmost vigor and with apparent effectiveness.[40] As a result of these policies, the index of hourly wages has remained practically stable. This index, obtained by averaging hourly wage rates for skilled and non-skilled labor in mines, industries, and transportation, remained unchanged from 1933 to 1937; it rose by only one point in 1938, and by one additional point in 1941.[41] There are good reasons to believe that no major changes have occurred since that time.

Stabilization of wage rates was not equivalent to stabilization of wage income. Wage income increased as employment increased and as hours were lengthened. The Nazis have, however, also restricted wage income. In common with the Weimar Republic, they raised social security dues and imposed a tax on wages.[42] They have further reduced wage incomes by collecting from every wage earner membership fees for the Labor Front and for the Strength through Joy organization; by imposing levies for the German Winter Relief Fund; and by urging workers to pay installments towards an inexpensive "People's Car" three years before the car was due to be delivered

[40] For details of these developments, see Hamburger, *How Nazi Germany Has Mobilized and Controlled Labor*, pp. 44 ff.

[41] According to the computations of the International Labour Office. See *International Labour Review*, August 1943, p. 261.

[42] Increased by 50 per cent in the higher brackets at the beginning of the war. See below our discussion of taxation of income.

(deliveries were never begun). The total amounts of these deductions are difficult to ascertain. Nazi estimates are too low, anti-Nazi estimates too high. But they have certainly been considerable. True, there was a return on some of the deductions; for their contributions to Strength through Joy, many workers went to the theater, on weekend hikes, and on cruises. But, as has been well observed, diversion of expenditure from food and clothing was predominant among the reasons for imposing these contributions.[43]

Finally, efforts have been made to restrict the spending of wage income. Campaigns were organized and devices invented to promote savings from wages. These efforts culminated in the enactment of the so-called Iron Savings scheme (law of October 30, and an enforcement decree of November 10, 1941). Under this scheme saving was made easy and attractive. The worker is no longer required periodically to stand in line before the teller's window in an understaffed savings bank; the employer makes the deposit for the worker. Such amounts as the worker wishes to save[44] are deducted from his total wage and deposited in an account opened in his name in a local credit institution. Moreover, deductions thus agreed to are exempt from the wage tax and social security contributions. On the other hand, the accounts are blocked for the duration, except by special permission in cases of personal emergencies and for extraordinary expenses. After about one year of operation it was reported that 3.5 million wage and salary earners had saved roughly 76 million RM per month under the Iron Savings Scheme.[45]

Frequently, these wage policies have been described as revealing the anti-labor or pro-capital character of the Hitler dictatorship. Whatever its character may be, it must be clearly understood that from the outset Nazi wage policy—in common with every other phase of Nazi economic policy—was part of a rearmament and war production program. The Nazis had decided to restrict production and importation of consumers goods in favor of war goods. At the same time they were firmly resolved to keep prices reasonably stable.

[43] Juergen Kuczynski, *The Condition of the Workers in Great Britain, Germany, and the Soviet Union, 1932-38* (1939), pp. 60 ff.

[44] Only specified uniform amounts, not percentages, are permissible. They range from a minimum of 0.20 RM on daily wages to a maximum of 39 RM on monthly wages. The intent obviously is to reduce computing operations to a minimum.

[45] *Das Reich*, Nov. 8, 1942. The estimated total of wage and salary earners eligible for Iron Saving at that time was roughly 20 millions. It can thus not be said that Iron Saving has been compulsory. In fact voluntary normal savings are absorbing so much excess purchasing power that the Nazi government can forego compulsion. Indeed, consumers' goods and services are now so completely rationed or so unavailable, and price control is so strict, that saving is practically the only outlet for excess earnings.

With the upward spurt in production since 1934 or 1935, requests for increased wage rates were received. Hitler and Schacht explained that higher wages would result in rising prices, and that to balance rising prices increased wages would, in turn, be necessary. Repeatedly they warned that higher wages would lead directly to the fateful spiral of inflation. The leading Nazis were fully aware that in a war economy stable wage rates were prerequisites for effective price control, which, in turn, was necessary to justify stabilization of wage rates.

Prices for farm products were increased, but retail prices were not allowed to rise proportionately.

Wage control was accompanied and followed by commodity price control. While wage control was exercised over every trade and industry as early as 1933 or 1934, commodity price control expanded gradually over a period of three years.

Farm products were the first to come under price control—in the fall of 1933. In this case two objectives were pursued jointly. The first was to stimulate production with a view to achieving a higher degree of national self-sufficiency. The second was to provide bounties to farmers who as a class were singled out as the mainstay and pillar of the regime. Rather than aiming at stable prices, the goal was first to raise prices paid to farmers.

In carrying out this policy the government acted through the marketing associations of the statutory Reichs Agricultural Corporation. Fixed, standard, or "base" prices were set periodically for the different grades and quality classifications of farm products, with suitable allowances for regional and seasonal variations, freight differentials, and other items. Products with a very wide range of grades and varieties, or those difficult to classify—such as cattle scheduled for slaughter, wool, leaf tobacco, and hops—were graded and classified by special committees in authorized markets.[46]

The success of this effort at control can be seen from the index of prices paid to farmers, which rose by a full 25 points in three years— from 77 for the agricultural year 1932-33 to 84 for 1933-34, to 94 for 1934-35, and to 102 for 1935-36 (1909-10 to 1913-14 = 100).[47] This was the index for *all* farm producers' prices. There were wide

---

[46] For more detailed information see Harry Lee Franklin, "Agricultural Price Control in Foreign Countries," *Foreign Agriculture*, 1939, pp. 50-55 and "Wartime Agricultural and Food Control in Germany," *Foreign Agriculture*, 1940, p. 186.
[47] *Statistisches Jahrbuch für das Deutsche Reich*, 1938, p. 319.

variations in the several classifications. Thus, the index for slaughter animals rose by 29 points, that for sheep by 45 points, cereals by 8 points, and wheat only 1 point. These variations show that regulation of agricultural prices was not intended to bring about flat increases. A major object was to stimulate the production of specified products. Direct production control has been made unnecessary, except to some extent in the case of bread grains.

With prices to farmers raised and wage rates held constant, the Nazis hesitated to pass on the full increase in the cost of farm products to the consumer. As an alternative they reduced the middleman's margin. Accordingly, fixed prices to farmers were supplemented by what amounted to maximum prices to processors, wholesalers, and retailers. As a result of this policy, the index of wholesale agricultural prices rose considerably less than the index of prices paid to farmers. The increase over three years was only 18.1 points—86.8 in 1933 to 104.9 in 1936 (1913 = 100). Because of control of retailers' margins, the increase in the cost of food for a worker's family of five was still smaller. It amounted to only 9.1 points—113.3 in 1933 to 122.4 in 1936 (1913-14 = 100).[48] Though the cost of food doubtless rose somewhat more than the official German index indicates, it is nevertheless clear that a substantial part of the increased prices paid to farmers was absorbed by the middlemen and the processors.

Since 1934 agricultural prices have been completely regulated at every stage, from production to the final consumer. In November 1936 the elaborate system of controls was integrated in overall price control under the Commissioner for Price Formation. In 1937-38 producers' prices oscillated slightly above parity. By November 1942 the index of producers' prices had reached a high of 112, but again the increase was not fully passed on to the consumer.[49] It should be noted that, as purchasing power increased and the supply of farm goods from abroad shrank, farmers' prices, originally regarded as minimum prices, in effect became maximum prices—a development similar to that observed for wage rates.

Having thus deliberately increased—within definite limits—the prices for farm products, curbing the prices of other commodities became even more essential. Prices of two groups of goods seemed particularly to demand attention: (1) domestically-produced raw materials and semi-finished goods; and (2) imported goods.

[48] Data from the same, pp. 317, 331.
[49] *Wirtschaft und Statistik*, 1942, No. 12, p. 415.

*The government effectively controlled prices of non-agricultural
domestic raw materials and semi-finished goods.*

Prices for the first group were set under cartel agreements as in
the past. In its early years, the Nazi government was favorably disposed toward cartels. As cartels constitute readily available instruments of control, it was a Nazi policy to control rather than to destroy
them. Indeed, in 1933 and 1934, and again in 1936, what was called
a "cartel wave" swept German business. Under an act of July 15,
1933, the government actually helped to rebuild cartels that had
broken down under the weight of the depression.

To many anti-Nazi radicals these policies appeared as added proof
that the Nazi government was merely an executive arm of "big
business." It should therefore be pointed out that the Nazi government did not promote cartels unconditionally. Rather, it used its aid
to cartels as a technique of price control. Frequently the Ministry of
Economy required a cartel to decrease prices, or not to increase them,
as a condition to forcing outsiders to join or preventing outside
underselling. Thus, beginning in the spring of 1934, through the
instrumentality of cartels, prices were reduced for key products such
as coal, aluminum, nitrogen, concrete, glass, and bricks.

It is nonetheless true that cartels tended to raise prices. No sooner,
however, was this trend visible than the government stepped in. A
price commissioner was appointed in the fall of 1934 who kept a close
eye on cartels.[50] For example, decrees of November 12 and December
11, 1934 required the approval of the commissioner before prices
under existing cartel agreements could be increased and also his con-
sent to conclude new cartel agreements. Similar restrictions were made
applicable to prices of branded articles, practically all of which were
merchandised under resale price maintenance agreements. With a
view to reinforcing these measures, agreements to regulate cost elements out of which prices are computed (Kalkulationskartelle) rather
than agreements to regulate prices were curbed by the decree of November 15, 1934. These restrictions proved successful. While by the
end of 1936 cartelization had developed to a point where it included
all domestically-produced raw materials, all semi-finished goods, and
at least half of industrial finished goods,[51] the index of wholesale prices

[50] The rather limited functions of this commissioner were transferred to the Ministry of
Economy in the summer of 1935. The 1934 commissioner must not be confused with the
Commissioner for Price Formation, appointed in the fall of 1936.
[51] According to official estimates. See *Weekly Report of the German Institute for Business
Research,* Nos. 49-52, Dec. 16, 1936, p. 104.

of industrial raw materials and semi-finished goods rose by only 5.6 points in these three years—from 88.4 in 1933 to 94 in 1936 (1913 = 100).[52]

In November 1936 cartel prices, along with agricultural prices, came under the control of the Commissioner for Price Formation. They were further curbed under a decree of November 23, 1940, which went so far as to outlaw informal price agreements and mere recommendations as to prices, unless approved by the Commissioner.[53] Under a decree of July 27, 1942, an overall cut in cartel prices was ordered. At present the cartels are in a period of marked decline. With the increasingly severe restrictions on the production of consumers' goods, many cartels lost current significance. Furthermore, government regulation of prices has been perfected to a point where apparently the authorities no longer need rely on the facilities originally offered by these groups. Nor is there any room in the Nazi war economy for restrictive cartel policies such as the penalizing of production or delivery in excess of agreed quotas. In the fall of 1942 it was reported that 1,000 out of a total of 2,500 registered cartels existed only on paper.[54] In the spring of 1943 it was forecast that only 500 cartels would survive. In the late summer a high authority announced that 90 per cent of all cartels had been or were being dissolved: "The time of cartels has run its course."[55]

The maintenance of a stable reichsmark was
essential to the control of import prices.

Experience in the early twenties had clearly shown that a progressive depreciation of the mark in the foreign exchanges was quickly followed by progressive increases in the prices of imported commodities in terms of German currency. In fact, the great German inflation stemmed largely from the depreciation of the mark abroad. Every increase in the cost of imported foodstuffs and raw materials was soon followed by compensating increases in wage rates, and this rise in costs necessitated still further increases in prices. Accordingly, the maintenance of a stable reichsmark was regarded by Dr. Schacht as indispensable to the prevention of inflation.

[52] *Statistisches Jahrbuch*, 1938, p. 317.
[53] For details of this legislation, see Louis Domeratzky, "Price Control in Germany—Policy and Technique," U. S. Department of Commerce, *International Reference Service No. 19*, April 1941, p. 10.
[54] *Hamburger Fremdenblatt*, Nov. 4, 1942.
[55] "Die Bereinigung im Kartellwesen," *Deutsche Bergwerkszeitung*, Apr. 21, 1943. Hans Kehrl, "Das Ende der Kartelle," *Das Reich*, Aug. 29, 1943. Some of the functions of dissolved cartels have been or are being transferred to the proper statutory trade associations or Reichs Associations.

The pegging of the German exchanges prevented the *general* rise in the prices of imported commodities that would have resulted from a depreciation of the reichsmark.[56] It still remained possible, however, for the prices of some imported commodities to rise because of market conditions outside Germany. Hence, direct control over the prices of such commodities was deemed necessary.

Prices of imported materials were regulated through control of commercial margins.

By 1934 the prices of raw materials on the world market had passed the low mark of the depression and were picking up again. A certain amount of pressure on the German price level due to imports was unavoidable. But it was possible to minimize the pressure by controlling the mechanism through which foreign prices were transmitted to domestic prices.

The Nazi method was to limit the prices importers could charge in German markets. The restrictions varied at the three successive stages of regulation. At the outset it was decided that domestic prices for certain imported goods should be permitted to rise *only to the extent* the price had risen on the foreign market. This was the rule laid down for textiles and leather under decrees of April 19 and 20, 1934, and for non-precious metals (except iron) under a decree of July 31, 1934. For these goods the commercial margin, expressed in absolute amount and not as a percentage, was frozen. The maximum domestic price could not exceed the cost of replacements on foreign markets of units of the same kind and quality, plus the commercial margin obtaining in a specified base period (Anhaenge-Kalkulation). Thus an increase in the absolute amount of the margin could not occur even though prices increased.

Less rigid limits were set under a decree of September 22, 1934, which applied to the bulk of imported raw materials. Under this decree the government allowed the *usual* costs and the *usual* profit margins. On application, the Commodity Control Boards which were then established set the domestic maximum price authorized by this decree. As usual costs and the usual profit margin are ordinarily computed in percentages of the value of the goods, the absolute price increase in the German markets could exceed the actual price increase in the foreign market. That is to say, the relationship between the

---

[56] As Germany did not freely convert the reichsmark into foreign currency, determination of its value through the unrestricted interplay of the foreign exchanges was avoided. To a small extent only was the reichsmark accepted abroad. Foreign trade was conducted principally under various forms of barter agreements.

foreign price and the domestic price was a proportionate one. This legislation relaxed the rules that had applied previously. At the same time it outlawed speculative commercial gains.

The limits just described obtained when the Commissioner for Price Formation took office in November 1936. As is well known, a pronounced rise in prices on the world market, especially in the case of raw materials, occurred during the first part of 1937. Under these conditions, the Commissioner deemed it inadvisable to permit any domestic prices of imported goods to increase in the same proportion as foreign prices. He therefore decided to return to the original Nazi method of freezing commercial margins, expressed as specific amounts. A decree of July 15, 1937, and an enforcement decree of August 10, 1937, declared that prices of all imported goods should be computed on the basis of average cost and average profit margins prevailing for comparable transactions during 1936. Moreover, as a rule the actual purchase price superseded the replacement cost in the determination of the domestic price. This method of course could not always be followed by the Commissioner, who reserved authority to set margins as he saw fit.

In the first three years of the Hitler regime price regulation centered on the four fields analyzed. In effect, however, control reached much further. Controlling the prices of foodstuffs (after an initial deliberate increase), the Nazis controlled the principal component of the cost of living of the masses. Controlling wages, they controlled a basic cost element entering into the price of the vast majority of commodities and services. Controlling the prices of domestically produced agricultural and mineral raw materials and the prices of semi-finished products, they controlled additional basic cost elements affecting the prices of manufactured goods. Controlling importers' margins, they kept within reasonable bounds the impact of increasing prices in foreign markets on the domestic price structure.

In summary, by effective restraints brought to bear on essential forces operating from the cost side, a large measure of control was ensured over the whole price system. Prices were not stabilized, but the increases were controlled.

GENERAL PRICE CONTROL

In the course of 1936 the upward movement of prices continued, in some fields at a somewhat accelerated pace. The wholesale index for manufactured consumers' goods—the pricing of which, as we have seen, had been interfered with relatively little—rose by 5.4 points from January to December.[57] Consumers were concerned alike over rising prices, deterioration of quality not expressed in price, and the occasional disappearance from the market of certain goods apparently being hoarded in expectation of future price rises. The concern was strong especially among the wage-earning population. Rising employment had considerably increased total wage income; but, individually, many workers felt that, with wage rates stabilized at the low 1933 level, their share in what appeared to be a general recovery decreased or was inadequate. Moreover, the disastrous inflation experienced in the early twenties loomed in everybody's mind, and fear spread widely that Germany might be heading toward a similar catastrophe. This fear was enhanced by the announcement, in September 1936, of another huge expansion of the nation's economic militarization effort, the so-called second four-year plan.

In this situation the Nazis deemed it necessary to expand the coverage of price regulation. They decided to supplement the existing control over costs by the establishment of a maximum for practically every price. A decree of November 26, 1936 introduced general price control. Far from coming as a novelty, this move came as a completion or a sum of existing partial controls already operating on a wide scale.

General price control covered the wholesale and retail prices of practically all commodities. Included were the prices of second-hand articles such as cars, machine tools, and construction machines; of work horses;[58] and of Christmas trees which were later painstakingly classified in three price brackets depending on their height, with special allowance for tops of trees abnormally long in relation to the rest of the tree. In addition to commodities, price control extended to real estate prices, to rentals and leases, and to compensation for all services. Rates for services of utilities, insurance premiums, commissions, and fees of every conceivable type were included; even dues to football clubs and other clubs and societies, as well as indirect taxes imposed by municipalities were covered.[59]

Relatively few commodities and services were exempt. Among these were postal stamps for philatelic collections. It is reported that the prices of certain stamps that ten years ago sold over the post-office

[57] *Statistisches Jahrbuch,* 1938, p. 317.
[58] Under a ruling of the statutory Reichs Agricultural Corporation issued on Feb. 20, 1940, prices were entered on a kind of equestrian passport recording the commercial career of the animal.
[59] Leonhard Miksch, "Wie arbeitet die Preisaufsicht?" *Wirtschaftskurve* IV (1937), p. 305.

counter now fetch 100 times the original price, while valuable stamps are practically unobtainable. Shipping rates set by international agreement were excluded, as were also prices for luxuries such as riding horses, Persian rugs, and oil paintings, but prices of some luxuries were included later.

Interest rates were regulated independently. In January and February 1935 the government cut the interest rate from 6 per cent to 4.5 per cent on 8 billions of municipal bonds and mortgage bonds outstanding. Under an act of July 2, 1936, the rates on 3 or 4 billion RM of private mortgage loans were reduced to 5 per cent and 6 per cent. In 1936 and 1937 the rates on industrial bond issues were reduced from 6 per cent to 5 per cent, and in some cases (Krupp) to 4.5 per cent. Other mandatory conversions took place in 1940 and 1941. In addition, on new industrial bond issues standard rates were set and gradually lowered from 5 per cent to 4 per cent. Government control of the various bank rates and rates on savings has been almost equally complete since 1934 or 1935.

Regulation of security quotations was gradually brought under control beginning in 1936. Originally control extended to foreign securities only. At approximately the time of the decree of November 26, 1936, brokers were directed not to quote any such security whose price rose above the quotation obtaining on November 13, 1936. Quotations of domestic bonds, governmental, municipal, and mortgage bonds, were regulated later on. Stock quotations were the last to be regulated. At the beginning of the war a boom started on the stock market. The official index of stock prices rose from 100 in September 1939 (1924-26 = 100) to 160 in September 1941.[60] About that time bank loans for the purpose of buying stock were ordered discontinued. Beginning November 1, 1941 selling stock over the counter was outlawed; the Minister of Economy ruled that after that date all credit institutions must sell and buy stock exclusively through the Exchange at official prices. In the spring of 1942 actual maxima were fixed for the quotations of three leading stocks, Dye Trust and the two Siemens issues. Later, all quotations of May 12, 1942 were declared standard quotations, to be exceeded "only in joint operation"—whatever that may mean—with the official Stock Exchange Commissioners, and—apparently flexible—maxima were set on the quotations of more stock issues. Moreover, since June 1942 the government has manipulated quotations through a combined technique of registering

[60] *Wirtschaft und Statistik,* September 1942, p. 320.

stock and sales from government holdings.[61] As a result of these policies a measure of stabilization was reached; in the latter half of 1942 the index of stock prices oscillated around 155 and 156. Eventually, in May 1943, the fixing of maxima for the quotations of industrial bonds was threatened.

It can thus be seen that by now Nazi price control has become or is about to become virtually all-inclusive.

Except for prices of imports, interest rates, and security quotations, general price control has gone through two stages.

*The freezing of existing prices characterized
the first stage of general price control.*

Maximum prices were decreed on November 26, 1936, retroactive to October 17, 1936. No person or firm was to sell or deliver any commodity or to sell or supply any service at a price higher than had been charged on the latter date. In the absence of an actual transaction on October 17, the price fixed was that which would have been charged if there had been a transaction. For new products—books for example—the maximum price was that charged for a comparable product.

In the case of products with an entirely new use value or made predominantly from new—mostly domestic substitute or synthetic— raw materials, special regulations issued on November 8, 1940 provided that prices of these products should be determined in the same manner as they would have been determined on the base date. Ceilings were put on net prices rather than quoted prices; that is, allowances, discounts, or other price differentials applicable on the base date were taken into account. For real estate there was no adequate base price. It was therefore decided that the taxable value as established by the revenue authorities serve as a yardstick for fixing the maximum.

Two classifications of goods do not lend themselves to the comparative method of price fixing. The first was non-standard goods, such as building construction; in this field special regulations issued on June 16, 1939 set maxima for the various cost factors entering into the actual price.[62] The other class consisted of goods that in the nature of things did not have a market price, namely, ordnance and other government orders. Prices of such goods were fixed on a cost-plus basis. Originally each contracting firm was allowed its individual cost. This

---

[61] See below, p. 80.
[62] Leonhard Miksch, "Vom Preisstop zur Kostenkontrolle," *Wirtschaftskurve* III (1939), p. 252.

—the very denial of a fixed or maximum price—was found to encourage waste and, despite elaborate regulation of the actual computation of the cost, to involve the government in unnecessary expenditure. Accordingly in the spring of 1942 the system was changed. For the individual cost of each contracting firm was substituted "the cost of a well-run firm," (eines guten Unternehmens) and a uniform price was fixed for all orders concerned. The price was either uniform for the whole industry concerned (Einheitspreis), or the firms comprising an industry were classified in a number of groups according to plant location and other factors affecting cost. For each group the costs of a well-run firm were then established and a uniform price fixed (Gruppenpreise). In effect, uniform prices for government orders were maximum prices. They were intended to stimulate reduction of costs and to give the efficient firm a differential. These uniform prices, of course, had no meaning in the case of contracts with firms that were sole suppliers on certain government orders.

In general, value relationships rather than nominal prices, were frozen.[63] For example, it is lawful for an insurance company to raise premiums whenever and to the extent that risks have increased, or for a landlord to raise rent in proportion to improvements made upon the house or apartment rented. Conversely, deterioration of the quality of goods and services, or cutting of quantities sold, calls for proportionate price reductions. Needless to say, both the admissible increases and the required reductions of price have been difficult to enforce. On the one hand, there was a tendency to make minor changes in goods or services offered, to present them as improvements, and to raise the price disproportionately. On the other hand, amounts —but not prices—of goods have tended to shrink, and inferior grades have tended to be sold at prices originally charged for goods of a higher quality. It is true—at least in the commodity field—that the practice of raising prices by exaggerating minor improvements has been virtually barred, as a by-effect of the progressive standardization of types and styles of both consumer and capital goods. But the retention of a price that was subject to reduction because of a decline in the quality or quantity of the goods sold has remained a popular method of raising prices above permissible maxima.

Raising and lowering prices for a variety of objectives
mark the second stage of general price control.

The prices of October 17, 1936 were not intended to last indefinitely. Rather, freezing was supposed to be a more or less temporary meas-

[63] The same, p. 254.

ure which would make it safe for the government to embark upon price manipulation. Such manipulation constitutes the second stage of general price control. Characteristically, the administrator placed in charge was designated Commissioner for Price Formation. He received full authority over all price classifications thus far mentioned except wages, interest, and security prices. Wages and interest rates continued to be regulated by the Labor Trustees and the Reichsbank, respectively, while security quotations were to be regulated by or under the Ministry of Economy. The Economic High Command, created at this time, provided the necessary co-ordination. The Commissioner, supported by a network of regional offices attached to or part of local government authorities, has endeavored to keep the German price level reasonably stable. Through hundreds and thousands of orders and regulations, he has raised some prices above and lowered other prices below original maxima.

Raising of prices has served two different objectives: (1) The prices of certain goods were raised to promote broad economic policies. One policy was the stimulation of agricultural production. Such was the purpose of a 10 per cent increase in the prices of milk, butter, and cheese granted in March 1940, as well as of an increase in 1940 in prices for hog hides and producers' prices for vegetables. Similarly, prices of potatoes, hogs, and poultry were raised at the end of 1942. Another policy was the expansion of high-cost domestic production of materials essential in war; this was the purpose, for example, of the 1937 increase in the tariff on—and in the domestic price of—imported rubber. A third policy was the curbing of consumption. Thus, at the beginning of the war prices of wine and beer were increased by taxes.

(2) Prices of other goods were raised to make allowances for rising costs. While the overall price maxima, stabilization of wages, and government control of interest rates resulted in the stabilization of many cost items, some costs continued to change. Thus, rising prices for imported goods have been a factor tending toward higher costs even though they were not wholly reflected in domestic prices. Other increases in costs have resulted from the introduction of new manufacturing processes and the use of substitute or more expensive raw materials such as synthetic rubber or fiber. But it must not be assumed that the price authorities granted a price increase whenever an unavoidable increase occurred in a cost item, nor must it be assumed that they always granted increases in proportion to added costs. On the contrary, they have sought by several methods to prevent increasing cost elements from being fully reflected in prices.

One such device was the requirement that an increase in one cost item should be offset by a decrease in another cost. Invariably, the price authorities insisted on a reduction of total unit costs by means of improvements in technical processes and in administrative methods. Forced absorption of additional costs out of profit margins was another device. Thus, when costs increased as a result of rising prices for raw materials, they would permit an addition to price equal to only the absolute increments in costs of goods purchased. A similar practice was followed in the wholesale and retail fields. While granting price increases was often inevitable, the increases were carefully controlled and pyramiding has largely been avoided.

Lowering of prices has also served two different objectives: (1) prices of certain goods were lowered—as the prices of others were raised—to promote broad economic policies. Stimulation of agricultural production is again an outstanding example. This was the purpose of drastic cuts in the prices of synthetic and potash fertilizers ordered in the first months of 1937. Another major policy was encouragement of consumption of certain domestic raw materials. As an illustration, price reductions for synthetic silk and staple wool were ordered in the second half of 1937. A third policy was reduction of government expenditure. For this purpose the uniform maximum prices set on government orders in the spring of 1942 were cut in all brackets by roughly 5 per cent as of May 1, 1943. A fourth policy was transformation of farms and rural communities into auxiliary workshops to make parts of arms and ammunition. There are good reasons to believe that this was one of the objects of cuts in the prices for motors and other electrical equipment, which were ordered for the benefit of agriculture in the summer of 1939.

(2) Prices of other goods were lowered to make up for price increases that had been specifically ordered, or that could not be prevented. Two distinct techniques were used to compel price reductions which would offset price increases. Under the older technique the Commissioner would order specified firms or branches of business to reduce prices of certain articles to specified levels or within designated margins. Thus, prices of radio tubes, incandescent lamps, watches, and certain brands of other articles (through the statutory trade groups) were reduced by government fiat in the second half of 1937. Other examples include the lowering of prices for sewing machines and typewriters in the summer of 1939; of certain wholesale margins for textiles in April 1941; of aluminum, reduced by 4.5 per cent in

June 1941, after having been kept frozen for four years. These examples suggest that, to a considerable extent, the prices lowered by government order were those for consumers' goods.

The second and more recent technique dates from the beginning of the war. Up to that time a businessman was safe so long as he charged a price not exceeding the ceiling of October 17, 1936, or such maximum as had been set subsequently by the price formation authorities. He was not penalized if, under the lawful ceiling, his profits increased. Lower unit costs due to increased volume of business or reduction of individual cost items worked to his advantage. This was changed by an order issued on September 4, 1939, requiring that prices be set "according to the principles of an economy pledged to back the war." Should profits increase as a result of the war or as a result of cost reductions occurring during the war, it was made mandatory that the businessman—acting on his own initiative—should lower prices to a point where the excess margin would be cancelled.

During 1939 and 1940 this legislation, which had been very loosely drafted, appears to have remained almost a dead letter. But at the end of 1940 (decree of December 8), the Commissioner decided to enforce it. Immediately the problem of defining the limits of excess margins had to be faced; that is, lowering of prices was dependent on specifications of "adequate" profits. Since this specification amounted in effect to control of profits, we shall discuss it in its logical place. (See Section IX below.) Suffice it to say here that the Commissioner ordered prices, but not necessarily of all products of a given plant, to be lowered to a point where henceforth they would not yield profits in excess of the authorized maxima.

Reduction of prices under the 1941 scheme has not been enforced indiscriminately. Thus, on June 6, 1941 the Commissioner ruled that an investment to reduce costs may be more important than an immediate price reduction. In line with this ruling, instructions were issued to the effect that if in such a case a business gives specified guarantees for adequate allocation of excess profits, the requirement of a price reduction might be dropped. Indiscriminate price reductions may also counteract or make unavailable desirable methods of absorbing excess purchasing power. Obviously on this ground, it was forecast that for goods and services such as luxury perfumes and motion picture performances the Commissioner would insist upon surrender of excess profits, rather than on a reduction of prices.[64]

---

[64] *Die Deutsche Volkswirtschaft* No. 17 (1941), pp. 615 ff.

Price control appears to
have been successful.

The level of prices in Germany has risen more or less steadily, rather slowly from 1936 to 1939 and at a somewhat more accelerated pace since the outbreak of the war. Thus, the wholesale price index rose from 104.1 in 1936 to 106.9 in September 1939, and to 116.1 in June 1943. The trend of retail prices can be seen from the index of the cost of living which rose from 124.5 to 125.7 and to 139.4 in the same periods.[65] In other words, the increase was only 12 points for wholesale prices and 14.9 points for the cost of living.

These figures are subject to important qualifications. In the first place, deterioration of quality, as openly admitted by Nazi writers, is not adequately reflected in prices. This factor, however, must not be overestimated as the quality of essential goods such as electric current, most raw materials, and a great many agricultural products in the nature of things is unalterable. Second, a number of low-priced articles, while included in the indexes, are often not available in the market. In order to increase their earnings without exceeding price maxima, producers tend to concentrate on higher priced goods. The Commissioner has endeavored to check these tendencies by establishing for a number of industries minimum percentages of less expensive goods that must be produced. This was done, for example, for production of liquor and clothing in the spring of 1941. In the third place, there is a "black market." Ample evidence of its existence is found in the German daily papers, which at frequent intervals give prominent space to long-term, life, and even death sentences for those who happen to be caught. Although reliable statistics or estimates are in the nature of the case unavailable, it may safely be assumed that black market prices are high above official ceilings. A peculiar variety of— or a substitute for—the black market is the barter trade which apparently has spread in recent months and gives the government a good deal of trouble. For example, the butcher would not sell a pound of meat, nor would the cobbler mend a shoe, unless he were sure to receive a pair of gloves in addition to or instead of the maximum cash price. However, since the trade in the black and barter markets relates primarily to consumers' goods and since less and less consumers' goods are produced, the impact on the general price level is limited.

In the light of these observations, the German price indexes do not tell the entire story. But even if the wholesale index had risen from 1936 to the summer of 1943 by 14 points rather than by the official

[65] *Wirtschaft und Statistik*, October 1940, pp. 439 ff., July 1943, pp. 185 ff.

12, and the cost of living by 17 points rather than by the official 14.9, would such adjustment essentially change the appraisal of Nazi general price control? Considering that the period covered included the last three years of preparation for war and almost four years of actual war—nearly seven years of extreme pressure on prices—the increase was indeed a moderate one. This remains true whether adjusted or official indexes are used to measure price changes. Moreover, it must be remembered that prices of some goods were deliberately increased by the government.

It is true of course that prices were at the lowest level in 1933 rather than in 1936. Even when the increases that occurred during the first three years of price control—10.8 points for wholesale prices, and 6.5 points for the cost of living—are added to the figures as adjusted, the picture does not change materially. The total increase over ten years, then, was 24.8 points for wholesale prices and 23.5 for the cost of living. Such increases would not be considered abnormal for the upswing of an ordinary, old-fashioned, periodical business cycle. Indeed, the German wholesale price and cost-of-living indexes of the summer of 1943, although adjusted, were still well below the respective highs of 1929. In the aggregate, therefore, Nazi price control may be reckoned a success.

FISCAL POLICIES AND PRICE CONTROL

The measures discussed above have been in the fore of Nazi efforts to control prices. It would seem that the Nazis have been less concerned with fiscal policies. The early approach, as we have seen, was from the cost side. Price margins, too, were enforced in some fields, thus checking profiteering price advances. The stringent general price controls inaugurated in 1936 were in the main directed toward supplementing these measures by the establishment of a maximum for practically every price. While the Nazis always encouraged savings, it was not until the war years that much attention was given to fiscal policies as a means of holding down prices. Such policies received increasing emphasis after the Russian campaigns necessitated the ruthless curtailment of consumers' goods production.

But what is referred to in English-speaking countries as the "inflationary gap" has never been closed in Germany. That is to say, the volume of available consumer purchasing power has been considerably in excess of the volume of available consumer goods and services. The Treasury has not financed its full requirements through taxation and bond sales to individual investors and savings institutions. While available data do not permit a precise statement as to the extent to which the government has resorted to the commercial banks, it may be noted that the floating debt, practically all of which represents short-term obligations sold to the banks, has increased from 1.5 billion RM in March 1933, to 14.1 billions in December 1939, and to 98.9 billions in February 1943.[66] Meanwhile the commercial banks have also purchased huge blocks of long-term bonds.

Reports from Germany clearly indicate that the public has all along had large purchasing power. Thus, according to the German daily papers, the movies are over-crowded and other forms of entertainment, despite blackout and transportation difficulties, have flourished. There are also black markets. There has been a skyrocketing of prices for such unregulated commodities as stamps for philatelic collections. And there was a marked rise in the prices of stocks until gradually control of quotations was made effective.

Why did not this excess purchasing power—the so-called inflationary gap—result in a bidding up of prices all along the line? The answer is to be found in the other controls that have been developed by the German government. In addition to controlling costs and to setting a maximum for practically every price, the government specifies the items and limits the quantities of goods and services each buyer may buy. It has been said above that government allocation has facilitated control of the prices of materials. It should be borne in mind that, in addition to allocating materials, the Nazi government has increasingly restricted entry into business, regulated investments, rationed or withdrawn consumers' goods, restricted traveling, and so forth. Licenses, purchasing certificates, priority ratings, ration points, are money's passport to the market; without a permit money cannot buy. There is, as we have seen, a black market for certain consumers' goods, but it is not possible to use excess purchasing power in bidding up the *general* level of prices. Unauthorized demand, by and large, is powerless.

Under such circumstances, what becomes of the excess purchasing power? The bulk of it has gone into savings deposits, which have risen rapidly during the war period. As the war picture has darkened, increasing amounts of currency have been hoarded, as Mr. Funk has openly admitted in a radio broadcast from Berlin on June 28, 1943. A major lesson of German price control is: with effective control over costs, with a maximum set for practically every price, and with control over the quantities that may be bought, excess purchasing power has little chance to exert an influence on prices.

[66] League of Nations, *Statistical Yearbook* 1936-1937, p. 280. The same, *Monthly Bulletin of Statistics*, May 1943, p. 142.

VIII. Control of Cost Accounting

Our survey of price control has shown that the fixing of maxima for practically all prices in 1936 was preceded by regulation of essential cost elements and that subsequently price control was closely geared to changes in costs. Arbitrary increases and decreases made by the government were the only exceptions. In general, cost increases have been the basis for upward revision of prices, and cost decreases for downward revision. As costs are not necessarily what the businessman says or thinks they are, it was more or less inevitable that the Nazis should proceed from government control of prices to government supervision of cost accounting.

A prerequisite for government action in the field of cost accounting is that firms shall keep books. This was already customary for the larger firms, but it was the exception for smaller shops and for the farms. Neither of the latter groups were covered by the old commercial code of 1897. In order to overcome this first obstacle, the keeping of books was made mandatory for retail trade near the end of 1938. Similarly, the statutory Reichs Agricultural Corporation promoted bookkeeping by the farmers, who were supplied with proper forms. No less than 500 offices for the purpose of providing the necessary guidance and control were set up by the Corporation.

Standardization of accounts and accounting procedures has greatly facilitated control of costs.

In standardizing accounts and accounting procedure the Nazis built on essential spadework that had been done by the highly developed German science of business management, and on the experience of a number of cartels and trade associations that had introduced or recommended uniform accounting systems for their members.

The Nazis have standardized accounts and accounting procedures throughout industry and trade and have made them compulsory. A decree of the Minister of Economy issued on November 12, 1936 requested the National Divisions of Business to draw up rules and regulations and to make suggestions as to standardizing accounts and accounting. These and other endeavors were consolidated in the basic joint decree of the Minister of Economy and the Commissioner for Price Formation issued on November 11, 1937. A "Model Chart of Accounts" and "Plans of Accounts" were established. The former provides for the grouping of accounts under 10 main headings and 100 sub-headings that are uniform for all firms in industry and trade.

All accounts—potentially 1,000—of course are not utilized in all cases, but it is not permissible to change the headings, their order, their arrangement, or their definition. The Plans of Accounts are further subdivisions of the Model Chart; each plan is uniform for all the firms within one particular branch of trade or industry. The Model Chart as well as Plans of Accounts were progressively introduced from 1938 on. The degree of uniformity of accounts reached in subsequent years can be inferred from a statement in the *Frankfurter Zeitung* late in 1941. It was reported that the price formation authorities were equipped with "tables in which for each individual process the comparative cost of all firms in the branch is noted."[67]

Clearly, standardization of accounts and accounting techniques was not an end in itself. Rather, it was a means whereby the price formation authorities could more readily ascertain and interpret the cost structures of thousands of business units. Uniformity of accounting practices has greatly simplified the task. This uniformity also made it convenient for firms to compare the effectiveness of management, an exchange of experience strongly encouraged by the Nazi government as full capacity was reached.

Controls of cost elements and cost computations have been rigidly applied.

Control of actual cost computations has tended to keep costs as low as possible. To this end, the Nazis have prescribed the particular cost elements that are admissible in computing total unit costs, the manner in which certain costs should be computed, and the maximum amounts allowable for certain costs. A (third enforcement) decree of April 27, 1937, designed to determine the price of leather, regulated—among other things—the computation of the costs of inventories. Materials to be used in manufacture were to be valued at the price at which they were purchased, not at replacement costs. During 1937 and 1938 the price formation authorities extended this principle to many other industries.

Since 1937 labor costs have also been regulated. We have seen that until June 1938 the Labor Trustees fixed standard wage rates, not

---

[67] Quoted by Dr. H. W. Singer in "The German Economy V," *Economic Journal*, April 1942, p. 18. For details of the German system see H. W. Singer, *Standardized Accountancy in Germany*, Occasional Paper No. V of the National [British] Institute of Economic and Social Research, 1943. It should be noted, moreover, that forms and records, office techniques, and equipment have also been progressively standardized. This should also be viewed as a phase in the overall drive towards standardization of products and rationalization of technical processes. See pp. 31-32.

maximum wage rates, and that it was not unlawful for the employer to pay wages above standard. The price formation authorities, however, established the rule that expenditure representing payment of wages above standards was not to enter computation of labor costs. In other words, they made standard wages—not wages actually paid—the upper limit of labor costs allowed. Obviously, this was a circuitous way of attempting to freeze wage rates without establishing actual maxima.

The margin of labor costs permitted was narrowed further by a decree of November 15, 1938. This time the net effect was to limit the hours of work that could be counted in computing labor costs. Only such hours as are worked in accordance with the optimum standard of technical efficiency may be taken into account. Depreciation charges were regulated by the same decree. Depreciation can be charged only on equipment contributing to actual production, and then only to the extent that it is caused by actual wear and tear.

Originally the decree of November 15, 1938 applied to a limited field only. It applied to computations of costs in all cases where prices did not lend themselves to control through ceilings, and where for that reason effective control of cost computation was of preeminent importance for the proper determination of final price, namely, to government orders of goods that do not have a market price—which were figured on a "cost-plus" basis. In this field, above all others, the Nazis found it necessary not merely to regulate the magnitude of certain costs but also to prescribe what are and are not admissible costs.

Thus, the decree of November 15, 1938 has closely restricted the items or elements that are allowed in the computation of total unit costs. For example, costs of scientific experimentation and investigation must not be included, unless such research was carried out at the written request of the purchasing agency. Neither may income tax, donations, interest—on owned as well as on borrowed capital—nor entrepreneur's risk be charged as costs. These items must be borne out of profits. The decree draws what by German standards appears to be a new line of demarcation between costs and profits. Profits are defined as the amount sufficient to compensate for: (a) taxes and other necessary outlays not included in costs;[68] (b) interest on the capital—owned and borrowed—which is actually used by the enterprise and is necessary for its operation; and (c) the risks assumed by the entrepreneur. The last two items are rigidly controlled. The base on which

---

[68] Such as donations, but only to the extent they are "adequate."

interest allowed as a cost may be computed excludes assets such as dwelling houses and idle factories. Except when they definitely represent working capital, securities owned and bank balances are also excluded. In obtaining the base for the computation of interest, the assets mentioned above must first be deducted from total assets. From the remainder, depreciation allowances accrued up to the time of computation must be deducted. Any borrowed capital obtained free of interest must also be deducted. On the residue, only interest at the yield rate of German long-term government bonds is allowed. The return to the entrepreneur is determined in keeping with the conditions prevailing in various lines of industry or trade concerned; when necessary, in keeping with the conditions under which a particular order is carried out. While some leeway seems to be permitted, it is stated that in case of uneconomical management a claim may not be made to the full amount of profits as defined. In view of the strictness of these regulations, it is not surprising that in the decree profits are specifically labeled "computed profits" (kalkulatorischer Gewinn).

On January 16, 1939 a joint decree of the Ministry of Economy and Marshal Goering extended the application of the principles set forth in the decree of November 15, 1938 far beyond the field of government orders, especially to manufacturing industries. Again, it was ruled that a return to cover entrepreneurial risks should not be included in costs, and that depreciation should be computed in accordance with the former decree. With respect to interest, the 1939 decree is somewhat more liberal than the 1938 one; inclusion in costs of items previously disallowed is permitted in special cases. In these instances, however, the basis for the computation of interest is to be kept as narrow as that prescribed in the pricing of government orders. The applicability of the principles of the 1938 decree was extended by comprehensive legislation enacted in 1941, which was designed to reduce prices to a point where excess profits completely disappeared.

IX. Control of the Volume of Profits

In the preceding section it was stated that profits, as defined by the Nazi government, include amounts required to pay taxes. Obviously, profits so defined are gross profits. In practice the tax machinery has been a potent weapon in the control of net profits. In addition, price control and various types of contributions exacted from certain classes or groups of business enterprises have been used to hold down profit volumes.

The corporate income tax and other taxes applicable to business earnings were increased sharply.

Taxes on German business measured by net income were heavy in the last years of the depression. When Hitler seized power, a flat rate of 20 per cent was levied on corporate income. Reduced rates, however, obtained for limited liability companies with a capital not exceeding 50,000 RM. Personal income tax rates applicable to income derived from unincorporated business and other sources ranged from 10 per cent to 40 per cent, and a surtax of 5 per cent (crisis tax) applied to any part of individual incomes in excess of 8,000 RM.

Although the Nazis eased the burden somewhat by means of special techniques, they did not reduce these rates. In order to stimulate investment, profits spent for replacements and renewals in industry and agriculture were exempted from income tax and corporation tax by an act dated June 1, 1933. This provision was extended by an act of October 16, 1934 to include all expenditure for capital equipment with an estimated life of not over five years. Such expenditures could be deducted from the net income of the business enterprise as arrived at for tax purposes. The net effect was to postpone the incidence of the tax burden rather than to ease it, since short-lived equipment was written off in a single year and no margin was left to charge against earnings of later years. This allowance was discontinued at the end of 1937.

Indeed, the rates on net income have been increased. The act of October 16, 1934 had already set a rate of 50 per cent on the highest personal income brackets. An act of February 17, 1939 brought it up to 55 per cent. At the beginning of the war, under a decree of September 4, 1939, the Nazis increased the rates then existing for each bracket by not less than one half. This increase took the form of a surtax; incomes below 2,400 reichsmarks per annum were exempt, and no one was to pay taxes equal to more than 65 per cent of income. (Later on, the limit was raised to 67 per cent as of July 1, 1942.) The stiffness of these rates put non-corporate business at a disadvantage as compared with corporate business, and adjustments proved necessary. The problem was to reduce taxation of personal income derived from business without thereby increasing purchasing power for consumers' goods. This problem was solved by an act of August 20, 1941 which reduced taxes falling on un-withdrawn profits. One half of such profits were exempted from personal income tax, but in no case could the amount so exempted exceed 10 per cent of total profits.

In the case of the corporate income tax, the first action of the Nazis was to add to the types of business enterprises liable to full corporation tax. An act of October 16, 1934 withdrew the privileges and exemptions of municipal public utilities, mutual insurance associations, cooperative societies, and the like. In addition, the rates were raised progressively. An act of August 27, 1936 raised the general flat rate from 20 per cent to 25 per cent for 1936, and to 30 per cent for each year thereafter. Under the act of July 25, 1938, corporations making net profits in excess of 100,000 RM were singled out and subjected to discriminatory rates; the rates were 35 per cent on net income for 1938 and 40 per cent on net incomes for later years. The rates on corporate income exceeding 50,000 RM were increased under the act of August 20, 1941. The increase was to be one-eighth for 1941 and another eighth for 1942 based on the rates of 30 and 40 per cent above. Finally a decree of March 31, 1942 raised the rate for corporations with incomes exceeding 500,000 RM by an additional eighth. With these changes, the 1942 corporate tax became 30 per cent on incomes up to 50,000 RM; 37.5 per cent on incomes up to 100,000 RM; 50 per cent on incomes up to 500,000 RM; 55 per cent on incomes exceeding 500,000 RM.

In addition, the Nazis have controlled excess profits. First they used taxation. An act of March 20, 1939 imposed a tax on increases in personal as well as in corporate income, but excluding incomes from agriculture and forestry. The tax base was the amount by which net taxable income in 1938 exceeded net taxable income in 1937. The rate applicable to this increase was to be 15 per cent; on the excess of 1939 over 1938 income the rate was 30 per cent. This tax was actually collected only in 1939 on the excess of 1938 over 1937 incomes; it was discontinued by an act of August 21, 1940. The reasons for its repeal are not entirely clear. Among other things, the tax did not reach excess *profits* only; all increases in income were taxed. As a consequence, a large number of exemptions, deductions, and special adjustments had to be allowed, and assessment apparently presented numerous and burdensome administrative difficulties.

Later on, another method was used. Rather than taxing excess profits, it tended to forestall their coming into existence. It was the method of price control. We have seen that in the course of 1941 the Commissioner for Price Formation ordered non-agricultural prices to be lowered to a point where profits in excess of "adequate" margins disappeared. To set permissible maximum margins, a number of regulations were issued in the course of 1941, for industry on March 5, for commerce in April, for handicrafts in June, for the professions toward the end of 1941. With regard to industry, for example, the Commissioner prescribed bases (Richtpunkte) for determining adequate profits, which were to be computed as percentages on both necessary (not actual) invested capital and turnover. Yields allowed are not rigid; considerable leeway is left for negotiation between the firms and the price formation authorities; and yields are lower for high-cost producers than for low-cost producers, "thus compelling less efficient plants to improve their condition." Instead of prescribing specific methods of determining adequate profits for wholesale and retail firms, a prewar year—1938 as a rule—was selected as the base; that is, the profits of the base year were accepted as "adequate."[69]

Of course, it is not always possible to set prices in advance in such a way as to prevent profits from exceeding authorized maxima. Indeed, whether or not profits have exceeded a certain point cannot often be ascertained before the end of the business year. On the other hand, to reduce prices on completed transactions is difficult, if not impossible. Moreover, the government, as we have seen, in certain instances and under certain conditions preferred the accrual of excess profits to price reductions. The setting of maximum profit margins was therefore accompanied by provisions whereby profits in excess of maxima allowed were to be surrendered to the state. It should be noted that they were to be surrendered to the Commissioner of Price Formation, not to the revenue authorities. This choice of enforcement agency reminds us that the limitation of profits under the 1941 scheme was in fact incidental to the control of prices.

[69] For details, see Ernest Doblin, "The German 'Profit Stop' of 1941," *Social Research* (1942), pp. 374-75.

It would seem that the duty to surrender profits that had not been absorbed by price reductions was not strictly enforced. Perhaps the Commissioner for Price Formation was not well enough equipped to handle the administrative problem involved. However this may be, the 1941 scheme was later supplemented or superseded (decree of March 31, 1942) by a regular excess profits tax.[70] The tax is applicable to industrial and commercial firms only, not to agriculture, the pet child of the regime. The original tax base was the amount by which business profits in 1941 exceeded 150 per cent of the profits realized in 1938. Exemption was granted to firms with profits in 1941 of not over 30,000 RM. Regulations issued on March 28, 1943 enlarged the tax base to the amount by which income from business in 1942 exceeded 120 per cent of business income in 1938, and reduced the exemption for corporations to incomes below 20,000 RM. The rates for both years were 30 per cent for corporations and 25 per cent for unincorporated firms. These rates seem moderate, but it must be remembered that war profiteering has been very largely checked directly. Government contracts dominate industrial production and, as we have seen, on government contracts the computation of costs is strictly regulated and profits are kept within reasonable bounds.

Under this system the concept of profits has substantially changed. Very largely, profits have ceased to be the flexible residual margin between the attainable market price and the sum of what the businessman himself considers his costs. The relationship between price and profits has been reversed. Price no longer determines profit margins. Rather, profit margins are set as a factor in price. Profits are controlled as an integral part of price control. They have become a more or less rigid percentage added to the costs as construed by the government in order to ascertain the final price that may be charged.

Besides general measures to limit profits or to prevent excess profits, there have been several levies in certain restricted fields, which have reduced profits even further.

*Special levies on one business are used
to pay subsidies to another business.*

In addition to or in lieu of government subsidies to business, it has been a Nazi policy to have business subsidize business. Under this policy, the Nazis have imposed levies on selected groups or classes of

[70] Technically, the amounts raised are paid into accounts, the use of which will be settled by the Minister of Finance after the end of the war.

business enterprise. These exactions are not recoverable through higher prices; their effect is to reduce net profits available after all taxes and impositions.

Non-exporting industries and apparently also certain branches of commerce were the first groups of business enterprises subjected to these exactions. The Nazis have raised funds from them for the purpose of paying bounties to exporters unable otherwise to compete in the world market. While similar plans on a minor scale existed in pre-Hitler days and since 1934 in the automobile industry, this all-inclusive plan dates from about 1935. Great secrecy, however, has prevailed as to the total amounts raised and spent per annum and also as to the rates of "contributions" fixed for individual "contributors." On the basis of information obtained from well-informed sources, it appears that total amounts raised have ranged from 800 million reichsmarks in 1935 to well above 1 billion reichsmarks in later years. A competent observer like Mr. Hans E. Priester believed the rates of contributions levied ranged from 2 per cent to 6 per cent on turnover, varying with the several branches of industry.[71] These estimates indicate clearly that the exactions from non-exporting business have been considerable. Nazi exchange and price adjustments throughout Europe, together with the disappearance of the overseas market, have eliminated the need for export subsidies. In the absence of any information as to their discontinuance, however, it may be assumed that they are still levied. Once opened, the Nazis are not in the habit of cutting off a source of funds. Rather, when a particular need no longer exists, they siphon off the funds to other uses.

Funds to be used as subsidies have also been raised from certain branches of agriculture and agricultural industry. The price of processed milk had been kept relatively low. In 1934 a scheme was developed whereby producers of milk for consumption paid levies to compensate producers of milk used for processing. Under a similar scheme announced in 1937, sugar refineries, breweries, and wheat flour mills compensated rye flour mills. Although frequently altered, both schemes are known to have been in effect until 1941, and it is probable they have been continued. These are merely a few examples of the use of subsidies in this field. The importance of such schemes is indicated by the fact that in 1940 the organizations of the Reichs Agricultural Corporation spent 772 million RM on subsidies of these and related types. Of this amount, 187 million RM was

[71] *Das Deutsche Wirtschaftswunder* (1936), p. 232.

raised by the Corporation from its members, and 585 millions was supplied by the Reich Treasury from general funds.

In addition to the groups considered, active businesses have been assessed certain payments for the benefit of businesses that are inactive or working at low percentages of capacity. Maintenance and reconstruction of plants laid up or destroyed by military action have been taken care of by the Reich Treasury. But when plants have been closed or production reduced because of lack of raw materials, labor, electric power, and the like, the Nazis have charged various costs to the more fortunate firms. Under legislation of February 10, 1940 they set up a so-called National Maintenance Subsidy Scheme (Gemeinschaftshilfe der Deutschen Wirtschaft). Under this scheme the various statutory trade and transport groups impose assessments on fully operating member firms from which necessary payments for the maintenance of wholly or partially closed-down plants are covered. Similar arrangements have been made for handicrafts, processors and distributors of farm products, and for business units coming under the Reich Chamber of Culture. Flexibility in the operation of this scheme can be obtained by pooling the funds, which is a permissible practice. Data for rates imposed are not available, but when in the late winter and in the spring of 1943 thousands of retailing firms were closed down, the contributions payable by certain types of surviving retailers were increased considerably. Total funds raised amounted to 95.8 million RM by the end of 1942; industry alone contributed 76.6 millions.[72]

Measurement of the burden on individual business enterprises resulting from the several levies is impossible. But whatever trend exists is probably upward.[73]

In this study we do not compare profits earned during the Nazi regime with those of earlier periods. From 1933 or 1934 to the early war years German business was generally quite profitable, though in varying degrees. Certainly in this period profits, after all taxes and

[72] *Frankfurter Zeitung*, Apr. 15, 1943. A total of 9,100 firms received assistance under the scheme. Of total disbursements, 62.8 million RM went to closed-down firms, 0.9 millions to firms working at low percentages of capacity. Lately, the Reich has been reported to be contributing to the funds, in connection with the mass closing down of retail stores and craftsmen's shops, in the first half of 1943.

[73] The Nazis cut into profits in various other ways not included in our analysis. They have charged the costs of at least some of the controlling agencies to business itself. For example, under an order of Sept. 4, 1934 the Commodity Control Boards collect fees for their support from persons and firms subject to their jurisdiction. Similarly, the price formation authorities, under rules issued on Apr. 6, 1941, charge fees for approvals and other types of authorizations required under the various price regulations. A fee is charged even though the application is not granted. Although one cannot determine their incidence precisely, there is a strong presumption that these fees cannot be recouped through higher prices.

other exactions, reached higher levels than during the great depression. But these profits were of a peculiar brand. Any appraisal of German profits after 1933 is incomplete and misleading if it stops with a consideration of the returns shown in the income statements. For, where do returns "return" to under Nazi control of business? Do they go in the wallets of the investor who put up the capital or in the vaults of the company that did the actual investing? Or do they appear only on the books? What use can a German firm make of the profits it earns?

X. Control of the Allocation of Profits and Other Funds

Profits that are nominally allowed are not permitted to slip out of control. The government tracks down profits in the accounts of business; it takes charge of the accounts; it sees to it that to a large extent profits are employed only as it thinks suitable. The uses as well as the earning—the allocation as well as the volume—of profits has been controlled. In addition, funds that are not directly related to earnings of the accounting period immediately preceding have been made subject to government allocation. These controls have been carried out in two ways: negatively, by restricting or barring outlets; and affirmatively, by directing the funds into specific channels.

Dividends—especially cash payments—were
subjected to rather narrow restrictions.

The Nazis have barred some outlets partially, some completely. Among outlets partially blocked, the most important was payments to the stockholder. Under an act of December 4, 1934 the Nazis limited cash dividends to 6 per cent, and 8 per cent was permitted for the few companies that had paid more than 6 per cent in the preceding year. These maxima were not ceilings on dividends, technically speaking; it was not unlawful to declare and to pay higher total dividends. Rather, dividends in excess of these limits for cash distribution had to take specified forms. They were to be paid in the "loan stock" of the Golddiskontbank (a subsidiary of the Reichsbank) which invested them in government bonds; the interest on these bonds was also to be invested in government bonds. Held by the Golddiskontbank as trustee, securities thus purchased were blocked until final allotment to the stockholders.

It might be argued that profits earmarked for distribution were fully distributed and that under this legislation the outlet to the stockholders had been kept fully open. In actual fact, however, distribution was being narrowed down to the percentage indicated. The Nazis did not stop with the retention of the securities. Under a decree dated December 9, 1937 the government "took over" the whole fund. In lieu of their claims for the securities, the stockholders received non-interest-bearing tax certificates which could be used in paying taxes due in the fiscal years 1941-45. The loss to the stockholders was estimated by a leading German economic review[74] at 20 per cent, by

[74] *Der Deutsche Volkswirt*, Dec. 17, 1937, p. 508.

the German economist F. A. Pinkerneil at 23 per cent on original investment and accrued interest.[75]

From the outset in 1934, it appears that company executives disliked and distrusted the substitution of government-handled securities for cash payments. Some companies undoubtedly distributed cash under cover of loans to subsidiary companies, and in other circuitous ways. More generally, a policy of reinvesting profits rather than distributing dividends in excess of 6 or 8 per cent was adopted. In this manner, the maximum cash distribution tended to become a maximum total distribution. To the end of 1937 only 175 corporations were reported to have made contributions to the loan stock, and at that time the total stock did not exceed 90 million RM. Of this amount, the Reichsbank alone was responsible for the payment of 12 millions. Despite increasing prosperity, the amount of dividends contributed from 1937 to 1940 was equally low.

Though the outlet to the stockholder for dividends in excess of 6 or 8 per cent tended to be closed, this did not occur despite the will of the Nazi government. By reinvesting profits rather than declaring larger dividends, the company executives thwarted the apparent objective of the law—investment in government securities. The actual result, however, effectuated another Nazi economic policy—the promotion of the self-financing of industrial expansion: reinvestment of profits rather than distribution to the stockholder of dividends in excess of 6 or 8 per cent was in line with the Nazi policy of plowing earnings back into business.[76] As a result of these policies the ratio of dividends to profits declined steadily. While in 1927 total dividends of German corporations were almost as large as aggregate amortization on invested capital, by 1938 dividends had shrunk to almost one-third of amortization charges.[77]

In time, payment of dividends in excess of 6 per cent was practically barred. The 1934 law was renewed in 1937, but it was allowed to expire in 1940. The "voluntary discipline" of business, which was then depended on to prevent declaration of surplus dividends, appears to have failed. At any rate, a number of companies declared dividends in excess of 8 per cent. The government countered with the decree of June 12, 1941, which restored the previous maximum rates of 6 and 8 per cent. But unlike the previous legislation, these rates were ceilings

[75] The same, p. 516.
[76] Although not every type of business, as will be seen below.
[77] *Die Deutsche Volkswirtschaft*, No. 23 (1941), p. 841.

on *total* distributions of dividends rather than on *cash* distributions. The loan stock arrangement was not renewed and payments in excess of 6 per cent were subjected to prohibitive taxation.

It is true that the decree of June 12, 1941 also opened up a substitute outlet for profits. Over-capitalized companies were authorized to increase their capital stock and to "distribute" profits by issuing stock dividends. Adjustments in the amount of capital stock could be made only within limits: (1) the Nazis released for distribution only such reserves exceeding 10 per cent of capital stock as had been accumulated to the end of the business year ending in 1938—in other words, profits realized at the peak of the rearmament boom were deliberately excluded from distribution; (2) funds on which the issues were based could not include hidden reserves; (3) the distribution of reserves was discouraged by placing a tax of 10 to 20 per cent on the face value of new stock issued. By the summer of 1943 the process of writing up the capital stock was practically concluded. Of 5,400 companies existing at the end of 1940, almost 1,300 had increased their aggregate capital stock of 9,027 billions by 48 per cent. In this manner new share capital in the amount of 4,380 billion RM was created.[78] As the total capital stock of German corporations before this revaluation was reported as 24 billions,[79] the increase was 18 per cent.

But any advantage gained from this procedure did not accrue necessarily to "big business," the supposed darling of the Nazi regime. Despite an expansion in terms of plants and facilities that must be truly enormous, the German Dye Trust raised its capital stock by the stock-dividend method only up to the level of the years 1926-31.[80] And some giants of the heavy industries like United Steel, the Kloeckner and the Mannesmann corporations did not declare any stock dividends under this plan.[81] In order to compensate for the recent stock dividends, companies that declared such dividends now tend to distribute less than the maximum of 6 per cent allowed under the 1941 legislation.

[78] *Frankfurter Zeitung*, Aug. 31, 1943.
[79] *National Zeitung* (Essen), June 11, 1942.
[80] *Deutsche Bergwerkszeitung*, July 12, 1942. In addition, 235 million RM were raised by a genuine increase in capital stock.
[81] *Frankfurter Zeitung*, Mar. 8, 1943. In a final appraisal this paper pointed out on Aug. 29, 1943 that the writing up had occurred largely in the case of large and medium corporations. It added, however, that in proportion to capital stock outstanding, stock dividends had been highest in the case of small firms; exceptions notwithstanding, the ratio decreased as stock capital increased.

Other employments for funds were
blocked completely.

The Nazis have not stopped with a partial blocking of the outlet to stockholders. They have also in effect barred the investment of undistributed profits and other balances in a number of fields where they could be employed to advantage. Since the early days of the regime, as we have pointed out, the Nazis have systematically curtailed consumers' goods industries and, more generally, activities not contributing to the rearmament and war effort. To the extent that the establishment of new business enterprises and the expansion of plants have been blocked, opportunities for the investment of undistributed profits and other surplus capital have vanished. The several techniques used to this end have been described previously, though from a different point of view. Undesirable investment has been checked directly through total prohibition or restricted licensing of new business enterprises and additions to existing facilities. It has been checked indirectly through the restrictions on the raising of capital as well as through government allocation of materials, equipment, and labor.

The government promoted the plowing back of earnings
and investment of surplus funds in other businesses.

While closing or narrowing certain channels, the Nazis, conversely, have directed profits and other funds into such other channels as they thought fit.

The first channel led back to the source; the Nazis have caused large amounts of funds to be reinvested in the business enterprise which earned them. As we have seen, the technique used was stimulation—through tax machinery and through limitation of cash dividends. But compulsion—direct as well as indirect—has also been employed. The Nazis are known frequently to have compelled plowing back by direct government order. An example of indirect compulsion was the embargo on the issuance of securities. The immediate purpose of the embargo, as has been emphasized, was to preserve the resources of the capital market for government issues and such private issues as the government deemed imperative. To the extent that it cut off firms from outside equity and loan capital, it compelled them to supply their own capital out of earnings. Another use of the indirect technique took the form of a threat to refuse subsidies for expansion unless the firm threatened was ready to reinvest a certain portion of its available funds.

The volume of funds that the Nazis, by applying various techniques, have caused to be reinvested in any year cannot be estimated closely. No doubt it has been very large. We know that the marked industrial expansion that has occurred since 1933 was to a very large extent "self-financed." This applies not only to physical plant expansion, but also to secondary features such as the training programs designed to provide trained man power to operate the expanding industrial plant.

It should not be assumed that under the Nazi system *self*-financing is synonymous with *independent* financing. In turn, self-financing has been government controlled. A German business enterprise is by no means free to allocate its funds to any kind of project. The government has gone so far as to specify the channels into which available funds may flow; that is, they have determined actual types and items of plant expansion. This was done in part by direct order from the government, coupled when necessary with promises of subsidies or guarantees of adequate returns. Other techniques included prohibition or restricted licensing of plant expansion, and control of the supply of raw materials, equipment, electric current, and labor. For example, a firm intending to build workers' dwellings or a new office building would not get the required authorizations to build, to buy bricks or other materials, or to hire construction workers unless it undertook to expand or improve production facilities.

The second channel led from the source to other business enterprises. The Nazis have compelled certain industries to invest large amounts of profits and surplus capital in firms and industries in which they would not have necessarily invested of their own free will. The industries thus tapped and the industries and firms supplied have been referred to at various points in this study.[82] It should be emphasized, however, that the techniques used to compel investment of profits and surplus funds in other firms and industries were not of a routine character. The large and varying size of the investments involved and their relatively small number suggest that the methods used were adapted to individual cases.

The Nazis have acted very largely through representatives of the groups charged with such financing—informal bodies, or statutory trade associations, or chambers of commerce. The government met with these representatives and placed a project before them. The financial requirements involved and the ability of the industry to contribute to the project were then discussed by representatives of

[82] See especially pp. 23, 24.

the industry and government. In due course a lump sum would be fixed to be raised by all or certain members of the industry. There are not necessarily formal letters requiring a firm to make payments towards a project. The technique of compulsion used is much more subtle. The scheme is "agreed upon" and its execution is left to the group of business enterprises involved. Thus committed, it is impossible for any firm to back out.

Financing the war has been facilitated by forcing various funds into the Treasury.

The third channel—actually a network of channels—led to the Reich Treasury. The Nazis have caused business to invest large amounts of profits and surplus capital in government bonds. German business has been forced to acquire government bonds in two ways: by purchase and by exchange of stockholdings for bonds. In addition, certain idle funds were turned over to the revenue authorities directly.

The buying of bonds has been mandatory for a long while. It would seem that in the first years of the Nazi regime it was mandatory in the main for savings and banking institutions, social security agencies, and insurance companies. By 1938 or 1939 it appears to have become a rule for other types of German firms earning profits of substantial amount—with the exclusion presumably of farming. The technique of compulsion used is very simple. The German government does not appeal to the public to buy war bonds. Nor does it act through the medium of syndicates of banks to underwrite loan issues. Acting to some extent through the statutory trade groups, the Treasury issues bonds as the need arises and allots them to business.

This is what the Nazis style "noiseless" financing.[83] It is unquestionably noiseless in that all that is needed for the Treasury to place government bonds is an estimate of the volume of liquid funds available in what used to be the market, a few telephone calls, and a few letters. More to the point, this technique should be called *forced* financing. Indeed, the sale of bonds is noiseless because the purchasing is compulsory.

Unlike "purchasing" of government bonds, the technique of forcing business to exchange holdings of stock for government bonds is of recent origin. Since the beginning of the war, it has become increasingly difficult for many German firms to replace inventories and to undertake normal replacements of machinery. As a result, many firms became abnormally liquid and the problem arose as to where to direct

---

[83] Compare a highly instructive article signed "Ag" in *Wirtschaftskurve* II (1941), p. 153.

surplus funds. Because of fear of a currency inflation, surplus funds tended to find their way into shares rather than bonds. According to Nazi commentators, the purchases of stock were more or less indiscriminate and regardless of yield.

Originally the Nazis seem to have intended to curb only excessive trading in stocks; at the beginning of March 1941 Reichsminister Funk warned that speculative excesses in the stock market would not be tolerated. As the boom continued and the demands on the Treasury were stepped up by the Russian campaign, he decided to make available for government at least some of the funds that had been going into stock purchases. On September 26, 1941 Funk announced plans for a registration of shares purchased by industrial and other firms since the beginning of the war, thereby intimating ultimate liquidation of at least a part of these holdings. Registration began in the spring of 1942. Originally it was confined to holdings in excess of 100,000 RM value (on the basis of the official stock quotations of December 31, 1941) purchased after September 1, 1939. Later on the exemption was lowered. Anybody who adds to his stock holdings by purchase after February 25, 1943 and thereby reaches or exceeds at the end of any calendar month a total value of 50,000 RM for shares purchased during the war must register all holdings acquired after September 1, 1939. On June 9, 1942 the Nazi government issued a decree requiring the surrender of registered shares to the Reichsbank in exchange for interest-bearing treasury bonds. The exchange was to take place on the basis of stock quotations of December 31, 1941, but inasmuch as the treasury bonds were retained by the Reichsbank acting as trustee for the owner, the exchange was one in name only. The total amount of stock affected by this legislation was originally estimated at between 700 million and 1 billion RM. As a result of extended registration since February 1943, the total has doubtless increased. The Reichsbank is calling the shares in installments; in the summer of 1942 it was assumed that calls would be made particularly on industrial companies which bought up large stockholdings from profits earned since the beginning of the war.[84]

A special procedure was developed in the case of certain reserves.

[84] Though carrying out what might be called a confiscation of stockholdings, the Nazis have not thereby embarked upon nationalization of the corporations concerned, even in proportion to stock acquired. There is no indication that the Reichsbank is using, or planning to use, surrendered holdings to claim membership rights in company meetings or actually to engage in management of corporations. Stockholdings surrendered are serving another purpose. It was announced that they would be used solely to control stock quotations on the exchange and would be sold in limited quantities to small investors. While primarily intended to mobilize for the government surplus funds tied up in stockholdings, registration and surrender of stock, together with other techniques, have also been used to facilitate government manipulation of stock quotations.

Amortization funds which cannot be spent on normal replacements and renovations during the war have been siphoned directly into the Treasury. Under a decree of October 30, 1941 industrial firms may pay to the revenue authorities one half of the estimated value of depreciable capital assets as shown in the balance sheet for 1940. Commercial firms may pay 20 per cent of the estimated value of inventories as shown in the balance sheet for 1938—a year when inventories were well stocked. "May" in the Nazi language was a euphemism for "must"; firms were expected to pay. In November 1941 the industrial funds involved were estimated to run up to somewhat less than 10 billion RM.[85] In December of the same year it was estimated that commercial funds would add several more billions.[86] A total of 766 million RM had actually been paid by the end of August 1942.[87] Since this sum had not risen by May 1943, it appears that under the steadily increasing stress of war other requirements were given first call on accumulated reserves.

The significance of these payments must be fully grasped. Turning over the reserves to the revenue office was not equivalent to paying a tax: according to the law, the funds will be turned back. For each contributing firm the revenue office opens an account which is blocked for the duration, and the plan provides that the amounts involved shall be available for the purchase of capital equipment or to replace inventories after the war. Nor can these payments be said to represent forced loans, for within the accepted meaning of the term even a forced loan bears interest. The accounts with the German revenue authorities, however, yield nothing either directly or through conversion into government bonds. The payments to these accounts must therefore be characterized as a mandatory and entirely gratis farming out of industrial and commercial funds to the government. This was a new development in the technique of government allocation of surplus funds. Although directed by the government into specific channels, investment in all other cases had been remunerative, at least potentially. By the end of 1941, however, the Nazis had graduated from noiseless financing to costless financing.

This is how the Nazis have controlled the major financial aspects of business operation. It was shown previously that control of the conduct of a firm sets in when the firm springs into life. It will now be seen that control extends even to the phase of termination.

[85] *Die Deutsche Volkswirtschaft*, No. 32 (1941), p. 1318.
[86] The same, No. 35, p. 1406.
[87] *Frankfurter Zeitung*, Oct. 25, 1942.

XI. Control of Termination of Business

The Nazis have controlled exit from business. As in numerous other cases, this was done in both a restrictive and an affirmative way. They have closed down firms that wished to continue; they have kept in operation firms that wished to close.

ENFORCED CLOSING

Closing down has served a great variety of purposes, which have changed with the passing of time. Originally, businesses were closed in order to starve out or to damage political opponents or Jews. Later, by weeding out what were considered surplus middlemen, creation of privileged groups with a definite interest in the new regime was facilitated. As the supply of materials dwindled, the Nazis closed down plants in order to restrict civilian production or to concentrate both civilian and war production in the best-equipped plants. Lately, release of available man power for employment in essential industries, together with saving of fuel, power, office and other space, and related facilities, has become the dominant objective. To this end, the shops and stores and facilities of hundreds or thousands of peddlers, retailers, and craftsmen have been shut down progressively since 1938. To the same end, since the beginning of the war closing down has been extended to banks and bank branches, insurance companies, newspapers, magazines (both popular and technical), department stores, wholesale trading firms, and other businesses. Closings have been at an unusually high rate since the Stalingrad disaster in January 1943. In May 1943 it was announced that the big banks had to count on cancellation of fully one-third of their local branch offices (including offices closed in recent years).[88] Another illustration of the proportions of this action: in March 1943 there were 1,500 enterprises in the wholesale butter trade as compared with about 5,000 a few years before, and 755 distilleries as against 12,000. Thus, the economic foundations of a large proportion of the German middle classes have been all but destroyed.

The right to close businesses has also been of great value in enforcing regulations applicable to business; it has been used effectively as a threat to ensure compliance. The Commissioner for Price Formation has actually resorted to "economic annihilation" of a number of persons or firms charging prices in excess of the authorized maxima.

---

[88] *Deutsche Bergwerkszeitung*, May 22, 1943.

Techniques used in the closing down are as varied as the purposes pursued. Government licensing to control entry into business, which the Nazis extended to cover a very large number of trades and occupations, has played a prominent part. Licensing was not confined to new firms and individuals entering the various trades and occupations but was also applied to firms and persons that had been in business long before licenses were required. Indeed, the Nazis did not make provision for an established business to enjoy continued existence under a "grandfather clause"; they required an established business to be licensed like a newcomer. These comprehensive licensing arrangements made it relatively easy to compel the closing of a great many facilities and to remove men from established positions.

Other devices used to achieve the same objective included (1) drafting of essential employees by the employment authorities; (2) refusal of revenue authorities to permit postponing payment of income and other taxes, a fairly general practice under normal circumstances; (3) blocking of supplies of raw materials or equipment by the proper control authorities; and (4) straight orders from the Ministry of Economy, acting mostly through the statutory trade associations. This device has been prominent in the wholesale closing down of plants to ensure total mobilization of available man power.

Using these and other devices the government is free to close down any firm at any time. Licensed firms or people fare no better than non-licensed ones, since the government may withdraw a license before it has expired. This was illustrated in a somewhat comical manner by the 1942 drive against "growing discourtesy"; the Ministry of Agriculture threatened hotel and restaurant owners with cancellation of licenses unless they improved their conduct toward the public.[89] In Germany there is no guarantee whatsoever that a firm, although perfectly willing and able, may actually continue to operate and exist. There is no social security for business: German business enterprise has no *right* to live. It is merely *authorized* to live—as long as it fits the plans of the government.

ENFORCED CONTINUANCE

Conversely, business has been forced to live. True, we cannot point to any specific cases where the Nazis have ordered firms or businessmen to stay on the job, but this is in no way surprising. Indeed, it

[89] Dispatch from Stockholm, *New York Times*, Apr. 11, 1942.

is quite natural that such cases should not be known, for whatever desire to cease operating exists cannot be expected to become vocal. It would be altogether useless for firms or businessmen to announce that they intended to close down or to retire. They know perfectly well that such announcements would gain them nothing but the ill will of the government and they are also aware that the government would immediately order them to return to the job. Fritz Thyssen had to flee Germany to escape his responsibilities.

No plant in Germany is free to close and no businessman is free to quit working without government authorization. Operation of a business must not be abandoned. This has been expressly stated in the case of a number of specified types of enterprise such as small and medium farms, the so-called "hereditary homesteads" (act of September 29, 1933); electric power plants (law of December 1935); and the mining industry (act of December 1, 1936). But continuance in business is not a matter of statutory enactment. It is an underlying principle of Nazi economy, which the above laws merely illustrate.

Technically speaking, it is not illegal for a businessman to retire in order to enjoy life or have time to follow an avocation, or for a firm to sell out because business is unprofitable. Nor is it unlawful for a businessman or firm to discontinue operations because they are fed up with government regulations. But it should be noted that, technically, strikes by the workers are also lawful. If the Nazis have not outlawed individual walk-outs any more than collective walk-outs, if they have not outlawed management walk-outs any more than worker walk-outs, the reason is that any abandonment of work is so out of line with Nazi principles and policies that legislative bans are not required. A businessman refusing to work would immediately be drafted under labor conscription, which under the order of June 22, 1938 applies to management as well as to workers. Under this law he would presumably be put back on his former job. It can thus be seen that operation of a business under the Nazi system, like entry into business, is a public service. Every business is a public utility. For the *right* to carry on business the Nazis have substituted the *duty* to carry on business.

XII. Obtaining Compliance with the Regulations

Surveying the various aspects of Nazi regulation of business, the question will be raised: By what methods was compliance with the regulations ensured? Perhaps more than on threats of jail sentences, concentration camps, elimination from business, and blocking supplies, the Nazis have relied on the device of organization. We refer specifically to the chambers of commerce and related groupings, as well as the trade associations.[90]

ADAPTATION OF BUSINESS GROUPS TO NAZI NEEDS

The Nazis have treated business organizations in the same ostensibly conservative way in which they have handled ownership and operation of business enterprise. They did not destroy the organizations (except where necessary in bringing about consolidations) any more than they divested the businessman of ownership and operation. Rather, they have used the organizations for Nazi purposes. That is to say, they reversed the direction of the pressure exercised by the organizations. From groups bringing pressure to bear upon government, they changed them into agencies of the government bringing pressure to bear upon business.

Preparatory to such action it was necessary for the Nazis to streamline the organizations as they existed in 1933. The network was straightened out and completed, the constituent organizations were strengthened, and the whole new structure was tied in with the government. To achieve these preliminary objectives, the Nazis proceeded systematically and rapidly. Whereas government regulatory agencies were set up haphazardly, business was organized on a definite uniform pattern. Whereas the creation of government agencies has been a continuous process, streamlining of the chambers and trade associations was completed in two years. Since 1935 only minor structural changes, consisting mainly of simplifications and consolidations, have been made. By the time regulations began to grow in numbers, and with an ever-increasing impact on business, the machinery necessary to ensure compliance was available and in good working order.

[90] Both types of organization existed in Germany prior to the Nazis. The chambers, established by statute, were agencies of business self-government, as well as pressure groups. The trade associations, operating under general legal authorization, were almost exclusively pressure groups.

Adapting the chambers to Nazi
requirements was a relatively easy task.

In 1933 there existed three kinds of chambers in Germany: Chambers of Agriculture, of Artisans, and of Industry and Commerce. At the end of 1933 the Chambers of Agriculture were suspended and merged into the new statutory Reichs Agricultural Corporation. On the other hand, the scope of the Chambers of Industry and Commerce was expanded. Small business, shopkeepers, and retailers were included under a Prussian state law of December 28, 1933, which was followed by similar legislation in the other German states. At the same time both the Chambers of Artisans and the Chambers of Industry and Commerce were stratified. Germany was divided into 23—later into 27—economic regions. For each region there was established an Economic Chamber (Wirtschaftskammer), with which were affiliated all the local chambers—both of Artisans and of Industry and Commerce, as well as the regional units of the statutory trade groups. The Economic Chambers, in turn, were affiliated with a new central Reich Economic Chamber.

Early in 1942 Germany had a total of 209 Chambers—Artisans, Industry and Commerce, and Economic Chambers. Recognizing that the number was excessive and tied up men and facilities that were more urgently needed for war requirements, Dr. Funk in the spring of 1942 issued decrees to consolidate the existing set-up. There was to be but one Economic Chamber for each Gau (Party District); it would absorb all the Chambers of Artisans, of Industry and Commerce, and the Economic Chambers located in the Gau. Actually, the process of absorption has been a slow one and may still be in progress. Apparently the small fish have resisted being swallowed up by the big fish. Former local or regional chambers tend to prolong their existence in the garb of local or regional branches or units of Gau Economic Chambers.

The trade associations were organized in four
sections, which embrace all business.

In the case of the trade associations, much more work was necessary in order to straighten out and complete the desired organization. To this end, the Nazis divided the whole field of German economy into four sections, each to consist of a number of associations.

First, a section was laid out for agriculture. The Reichs Agricultural Corporation (Reichsnaehrstand) established under legislation of September 19 and December 8, 1933 included all the individuals and organizations engaged in and concerned with the production, processing, and distribution of agricultural commodities. There are local chapters of members of the Corporation as well as some 500 district and 20 regional groupings. There were also thirteen[91] functional associations (Hauptvereinigungen, Wirtschaftliche Vereinigungen) that are specifically charged with regulating the various phases of the marketing of agricultural products.

Likewise toward the end of 1933 (act of September 22) a section of trade associations (called chambers) was laid out to cover the field of "culture" (Kultur). It embraced authorship, the press, the theater, music, painting and plastic arts, motion pictures, and broadcasting.[92] Each of the respective associations is subdivided regionally as well as functionally. Membership includes all persons engaged in the production, reproduction, distribution, custody, and sale of "cultural products," very broadly defined, and covers a great number of business enterprises, in addition to persons belonging more strictly to professions. The added chambers form a Reichs Chamber of Culture.

Third, trade and industry were organized on Nazi lines. Under a decree of November 27, 1934 six National Divisions of Business (Reichsgruppen) were established respectively for industry proper, trade proper, the skilled crafts, banking, insurance, and power. A National Division for hotels and restaurants was added in 1939. The so-called trade groups (Wirtschaftsgruppen) are the backbone of six of the National Divisions, and indeed of the whole structure. On the other hand, the skilled crafts are lined up in some 17,000 "guilds" which in turn are members of some 50 central groupings. The number of trade groups has changed as a result of successive reshuffling. By the time the war began 31 groups had been carved out of industry and 15 out of all the other classifications.

Membership in the groups varies widely. At the beginning of the war the mining group had 50 members, the retail trade group 500,000. The internal structure of the groups varies accordingly. Many, but not all groups, break up into functional branches (Fachgruppen) and sub-branches (Fachuntergruppen). The Minister of Economy struggled continually to check the German passion for organization until, by the beginning of the war, he had reduced the total number of

---

[91] In the course of the recent consolidation drive some of the associations were merged, for example, under a decree of the Minister for Agriculture of Jan. 7, 1943.
[92] In October 1939 the broadcasting chamber was merged with the monopolistic Reichs Broadcasting Corporation.

branches and sub-branches from 400 to 328 and from 650 to 327, respectively. Likewise, many but not all trade groups, branches, and sub-branches form regional or district units that are affiliated with the Gau Economic Chambers. For example, half of the industry groups have been directed from the center in Berlin while the mass organizations of retail trade have been highly decentralized.[93] The whole set-up is topped by the Reichs Economic Chamber, which thus heads the network of both the regional chamber organizations and the functional organizations of trade and industry.

The last section of business to be organized by the Nazis was transportation. By decree of September 25, 1935 the Minister for Transport established seven functional Central Transportation Groups—one each for seagoing shipping, inland shipping, motor transport, carrier services, rail vehicles (streetcar lines, private railroads), forwarding and storing agencies, and auxiliary services such as sleeping cars. The transport organization has functional and regional subdivisions and is closely tied to the set-up of trade and industry.

These four sections and their constituent associations cover the entire field of business. The network is so tight that not infrequently "jurisdictional disputes" have developed between associations. In some cases the jurisdictional question was resolved by compelling a particular business to affiliate with several organizations. A textile manufacturer who has added a wholesale trade department to his plant and operates a canteen for his workers is bound to join—and pay membership fees to—the trade group for the textile industry as well as the groups for wholesale trade, and for hotels and restaurants. In other cases formal agreements were concluded between trade groups, or between the groups and the Reichs Agricultural Corporation or the Chamber of Culture, to draw lines between their respective spheres of interest.

Their monopolistic position and compulsory membership
greatly strengthened the business organizations.

Each of these organizations has a monopoly in its appointed field. This was not a new achievement for the Chambers of Artisans and the Chambers of Industry and Commerce. There had always been only one of each in every region or locality. Rival trade associations, on the other hand, existed in many fields. As a rule, under the Nazi

---

[93] Facts and figures in this paragraph from Eberhard Barth (a high official in the Ministry of Economy), *Wesen und Aufgaben der Organisation der gewerblichen Wirtschaft* (1939), p. 53.

scheme the strongest was granted a monopoly and the weaker ones were dissolved or swallowed up. Nor has the businessman any opportunity of building up new rival organizations, should he disagree with the policies or methods of the statutory ones. Not that such action has been outlawed by any enactment. Under the Nazi technique of repression nobody would even think of attempting such a thing.[94]

Compulsory membership is a second factor strengthening these organizations. For the Chambers of Artisans and of Industry and Commerce this condition also had existed prior to Hitler. For the trade associations, however, compulsory membership was a new achievement. Persons or firms who come under the Reichs Agricultural Corporation or the Reichs Economic Chamber are automatically members of the proper associations; in certain cases they merely have to report their existence. The Reichs Chamber of Culture, on the other hand, requires that membership be applied for, and an application may be rejected—with the result that the applicant is precluded from exercising his trade or vocation. There is thus in Nazi Germany a closed shop for business as well as for labor.[95]

The organizations as refashioned by the Nazis have become part of the governmental structure.

The organizations thus systematized and built up were closely tied to the government. The top of the new structure is actually a part of the government. The head of the Reichs Agricultural Corporation is the Minister for Agriculture, a post held by Mr. Backe since the spring of 1942. The head of the Reichs Chamber of Culture is the Minister for Propaganda, Mr. Goebbels. While the Minister of Economy and President of the Reichsbank is not formally head of the Reichs Economic Chamber, the latter is practically a part of the Ministry of Economy. Moreover, the Gau Economic Chambers— established in 1942—were put directly under the Minister of Economy.

[94] Similarly, as we have seen, there is no legislation in Germany to outlaw strikes and lockouts!

[95] This is not to say that membership is actually 100 per cent in all the statutory groupings. When the organizations were built up, they occasionally "passed the buck" on doubtful cases. A number of prospective registrants were sent around from office to office until they became exasperated, and decided not to register with any group—thereby saving membership fees, and at the same time avoiding other inconveniences. These cases came to light later on, when some of the trade groups were made responsible for the allocation or sub-allocation of raw materials and equipment. Then suddenly the slackers crept out of their retreats. (*Der Deutsche Volkswirt*, Nov. 15, 1940, p. 7.) For all practical purposes, however, it may be said that every German businessman or firm is affiliated with a statutory business group. If engaged in commerce or industry or a skilled craft, the German business has two affiliations—with a (regional) chamber and a trade association. Actually, as we have seen, a business is often a member of several trade associations.

The lower strata are tied to the government through the higher strata as well as through the operation of the leadership principle. The gist of this principle is that the government appoints the ranking leader of each organization and he in turn appoints the minor leaders. For example, the Minister of Economy appoints the head of the Reichs Economic Chamber, the heads of the National Divisions of Business, of the Economic Chambers, and the trade groups, while the heads of the National Divisions appoint the heads of the branches and sub-branches. The majority of these leaders are active businessmen. Even the appointment of an organization's managing director by the head of the organization is in certain cases subject to approval by the head of the next higher unit.

Advisory Councils have been established in a number of organizations. But their membership, too, is appointed not elected. To our knowledge they are hardly more than sounding boards for the leaders. Nor does the rigor of the leadership principle seem to have been weakened essentially by a recent attempt to provide the leaders with more worthwhile co-operation of their fellow businessmen. A decree of the Minister of Economy of January 30, 1943 proposes to dissolve the Advisory Councils existing in the trade groups and branches, and to replace them by Boards (Praesidien and Vorstaende) that apparently are intended to enlist the help of more active rather than of "yes" men for the conduct of some of the organizations.

It can thus be seen that in Nazi Germany every chamber and every trade association is controlled by the government, while in turn every businessman or firm is in the grip of one or more of these organizations. Through the instrumentality of organization, the government reaches down to the last businessman or firm in the remotest corner of the Reich.

PROPAGANDA, POLICING, AND REGULATORY ACTIVITIES

In the Nazi scheme of control of business the chambers and trade associations are the government's combined morale and enforcement department. In a system that keeps the businessman on the job, they endeavor to make sure that the job be done according to government policies and regulations.

To this end the organizations continuously lecture, circularize, instruct, and supervise business. Their action is both propagandistic and technical. They rub in the principles of the Nazi code of life, explain to business why it is being controlled and for what purpose, preach to business the necessity of putting up with the inconveniences involved in being regulated and require business to make sacrifices in the national interest.[96] In a word, the government uses the organizations as agencies of indoctrination and mental regimentation, thereby ensuring a fair measure of co-operation by business.

On the technical side, the organizations hand down and explain to business the meaning of specific regulations and they also tell business how to apply them. Through this work the organizations become well acquainted with the internal affairs of many of their member firms. As a result, they readily learn whether and to what extent regulations are actually being carried out, and see to it that they are enforced. That is to say, the government uses the business organizations as a new and unique type of industrial police.

It is not claimed that this police force is necessarily 100 per cent effective or even reliable. To be able to explain regulations, officials of organizations have to be well acquainted with regulations; they naturally come to know the loopholes and avenues of evasion better than anyone else. For example, we have it from an excellent German source that the officials have occasionally used their knowledge to point out to members how to get around certain measures of price control.[97]

To some extent the work of the organizations includes actual regulation of business.

The jurisdiction of the chambers and trade associations is not confined to ensuring compliance of their members with government policies and regulations. The organizations are closer to the field than the regulatory agencies of the government; they are naturally highly familiar with conditions in the branches or regions they represent. For these reasons they have been given a share in the actual regulation of business.

Some of the organizations are consulted when a person wants to go into business. In the most recent phase (spring 1943) of the drive to close down retail stores, department stores, shops of skilled craftsmen, and other business, the local units of the trade associations or the local branches of the chambers—as the case may be—made preliminary recommendations as to which specific shops should be closed. Probably they had similar functions in prior phases. The trade groups

---

[96] Occasionally, this task of the organizations is set forth in an extraordinarily naive manner. As long ago as 1938 a Nazi author explained that the Reichs Agricultural Corporation was providing "mental support," (weltanschauliche Schulung) especially for distributors of farm products, since the commercial margin allocated to the latter had been severely cut by marketing regulations. Bernhard Mahrens, *Die Marktordnung des Reichsnaehrstandes* (1938), p. 17.
[97] Eberhard Barth, *Wesen und Aufgaben* (1939), p. 82.

and their subdivisions lay down rules requiring that certain technical methods of manufacture be used and contribute in various other ways to the progress of rationalization. They see to it that the adjustments necessary in using new synthetic materials are made in time. They prohibit the use of wrappings beyond a certain limit and provide that boxes and crates shall be returned to the sender rather than burned. The organizations issue regulations as to the proper use to be made of unavoidable waste materials, they decide whether or not rebates may be granted, and they fix rates for commissions.

Applications to the price formation authorities concerning price increases are cleared through the proper trade groups, which weed out requests that are ill-founded. Especially important are the responsibilities that have been entrusted in recent years to the trade associations and their ramifications or related groupings in connection with the allocation of raw materials. Some of the trade groups also contribute prominently to organizing the looting of German-occupied countries.

These examples could easily be expanded. As controls have broadened and stiffened, the administrative work required for their application has grown more detailed and complicated. In an effort to alleviate the burden on the regular government agencies, the Nazis have assigned or shifted an increasing part of it to organizations of business. The Reichs Agricultural Corporation, of course, exercised essential regulatory functions from the very beginning.

These trends should not be regarded as indicating the existence of business self-government.

The trends discussed have often been misinterpreted as meaning that German business was living under rules of its own—that it was autonomous or self-governing. It has even been suggested that through the organizations business—especially "big business"—was controlling the Nazi government rather than the Nazi government controlling business. This interpretation is due in part to the fact that many active businessmen have been kept or placed in charge of the organizations. Moreover, in an obvious attempt to compensate business for the loss of its former substantial independence, the Nazis insist on referring to both chambers and trade associations as instruments of self-government of business. Indeed, periodically, campaigns are started and reorganizations undertaken to promote what is called self-government.

In reality there is no self-government of business in Nazi Germany. Germany is not a corporate state, any more than Italy. After initial hesitation the Nazis dropped any such pretense. None of the business organizations formulates policy. The organizations act exclusively within the ever-tightening framework of instructions and directions received from the operating or co-ordinating agencies of the government. They complete these instructions and directions; they amplify them; and they apply them to individual cases. They assist rather than replace the agencies of the government. The same observation applies to the work of the organizations in allocating raw materials which is carried out under authority of the Ministry of Economy. The functions of the organizations in the process of directing the conduct of business enterprise are of an executory nature. Nor is there any evidence that by circuitous ways the organizations control the government. That they presumably make suggestions and express their wishes to the government is a different matter. Within the framework of Nazi ideas and institutions, anybody may—and many people do—offer suggestions and make known their wishes.

This condition is not surprising. Germany is a total state. The concepts of total state and of self-government are mutually exclusive. What in Germany is advertised as self-government is in reality state administration, carried out by non-governmental agencies. The more honest among the Nazis are quite frank about this situation when they do not write in the daily papers. Thus a leading Nazi official admitted not long ago that "today self-government [local or functional] and government by the state are no longer differentiated, except in respect to form."[98] Indeed, for a totalitarian government to leave certain administrative jobs to be done by a non-governmental rather than by a governmental agency is merely a matter of administrative convenience.

The Nazis have chosen the very organizations of business to direct —to some extent—the conduct of business enterprise. The shrewdness of this choice should not mislead us. If the organizations, their chairmen, and their officers drive wheels, they in turn are merely cogs in the government-driven machine of the Nazi state.

[98] Dr. Ehrensperger, *Reichsverwaltungsblatt* No. 13-14. Quoted from August Dresbach, "Kaufmann und Behoerde," *Wirtschaftskurve* II (1942), p. 103.

XIII. Evasion of Control and Slowness of Control Action

To some extent the operation of the Nazi statutory organizations of business is thwarted by forces inherent in the system itself. Pitted against the efforts to ensure compliance with regulations are tendencies to soften up regulations. Pitted against the efforts to ensure an adequate administration of controls are disorganizing forces that result from over-regulation.

"Fixing" and graft lessen the
effectiveness of control.

Under the Nazi system of comprehensive control of business, the bulk of regulations is extremely elastic, and the operating agencies have received wide leeway in applying and enforcing regulations. They quite generally have authority to make special provisions or to grant exceptions from general regulations, when a case seems to warrant special consideration. This leeway, of course, is intended to ensure that control be adjusted to the peculiarities of individual cases. It is intended to, and certainly does, enhance the effectiveness of control. But it also works in reverse. It opens up an avenue of escape from the full incidence of control. It affords the German businessman opportunities for bringing about interpretations and applications of rules and regulations designed to promote his own interests rather than the policies of the government.

To make the most of this opportunity, it is necessary that the businessman be on excellent terms with the controlling agencies. In the absence of an articulate public opinion, a parliament, and organized pressure groups, the German businessman has no need for a public relations department. To influence the powerful agencies of control, however, he has good use for what might suitably be called a private relations department. Under the Nazi system of control of business by an absolute government, the contact man or graft, or both, take the place of the public relations executive.

The contact man is primarily a political figure. His job is to pull wires. He knows influential members of the all-pervading Nazi party in a position to bring pressure successfully to bear upon the men in charge of controlling agencies. Or, better still, he is an influential party man himself. Two types of contact men are known to be used: one an independent agent whom the businessman hires—or attempts to hire—whenever necessary; the other carried on the pay roll of the business in a more or less permanent capacity. Contact men of the second type are employed either under cover of some executive position or are openly appointed to the Board of Directors. Frequently such appointments were made by corporations to replace Jews, politically suspect board members, or members whose terms expired. In other cases, boards or executive bodies were enlarged in order to create seats or functions for new members with party "pull." It was not until the spring of 1943 that an order was issued to cancel existing and to prohibit future appointments of this kind.[99] But it applies only to full-time officials of the Nazi party and to members of the so-called Reichstag, nor does it outlaw appointments to "facilitating" jobs.

Under the Nazi system the use of a contact man clearly borders on graft, and many a contact man is but a thinly-veiled agent of corruption. But if and when the services of a contact man are not available, are not considered expedient, or fail to achieve the objectives desired, it is not unusual for the German businessman to pay government or party officials straight bribes. True, little official evidence is available to show the existence of graft in Nazi Germany; the cases reported as brought to trial are not numerous. Indeed, it would seem that the Nazi government deliberately tolerates a certain amount of corruption; apparently it considers moneys thus paid and received as a kind of toll on business levied in favor of deserving party members and other pillars of the regime. It appears to accept graft as more or less due the Nazis for having left the businessman on the job and for providing a steady flow of orders. In any case, the testimony of businessmen and other competent observers who have done business or observed business methods in Germany since 1933 is so much in agreement that the widespread existence of graft cannot be doubted.[100]

But the Nazi system of business control does not lead merely to corruption and favoritism. It is also responsible for a tendency of German business to operate slowly and cumbersomely.

Abundance of regulations creates
confusion and delays.

First, there is the truly amazing number of regulations emitted by the control agencies. An unending stream of enactments, decrees,

---

[99] *Das Reich*, Apr. 18, 1943.
[100] See, for example, the colorful account given by Douglas Miller, *You Can't Do Business with Hitler*, pp. 198 ff.

rules, circulars, instructions, and the like pours forth from the offices of the economic high command and the operating agencies, as well as from those of the statutory business organizations. Many agencies even publish gazettes of their own, issue upon issue, in addition to the regular government publications.

So great is the output of regulations that the businessman often is completely in a quandary as to what he should do. Which agency regulates this specific action of mine—Agency A, Agency B, or both? Do I come under Instruction 12, 15, or 23? If under the latter—is it still valid or has it been superseded or amended? If amended, is the changed version not in contradiction to an overruling order received last week? If not in contradiction—how am I to interpret it? What does it really provide for? What on earth does the government want me to do? Am I safe in taking this line of action, or will it land me in jail?

As has been well stated: "In the place of a wisely limited supply [of laws] we have a plethora such that one might almost speak of an inflation of labor legislation. This inflation has burst upon both employers and employees, making it practically impossible for them to find their way through the rules and regulations under which they are supposed to live."[101] This statement, made by one who took a leading part in shaping these regulations, bears apparently on conditions relating to labor control. With equal accuracy it pictures the quantity of regulations to control business, as well as the resulting uncertainties in the conduct of business enterprise.

Uncertainties or no uncertainties, the businessman must take action eventually. Ordinarily he does not act until he has made a serious attempt to clarify his position. The length of delays in the conduct of affairs varies of course. In some instances, the businessman is able to clarify his problem by referring to loose-leaf editions of laws and regulations which have become increasingly popular as the number of regulations has grown. These services keep track of regulations as they are issued, itemize them, and to some extent provide explanations. To keep up to date and to consult such services does not slow down action considerably. Not always, however, can clarification be obtained so readily from the explanations available. Frequently a more protracted analysis is required which may cause much loss of time. This is especially true when the businessman refers his problem to lawyers, either directly or through the channels of the proper statu-

[101] Werner Mansfeld, *Deutsches Arbeitsrecht*, September 1942, p. 117.

tory business organizations. "The attorney's office is my torture chamber," a well-known German banker once stated to the author. Consultations with lawyers are necessarily frequent. Indeed, it would seem that under government control German business has become a happy hunting ground for attorneys. With the conduct of business curbed and directed by the government, opportunities to draft the terms of contracts and to carry on actual litigation apparently have greatly decreased. But it would seem that a vast new field for lawyers has opened up in the analysis of the ever-growing number of regulations and in related negotiations with government agencies.

Bureaucracy slows up action.

In addition to their number, it is the administration of regulations that causes German business to operate slowly and cumbersomely. In applying regulations, the agencies of the German government tend to entangle business in red tape.

Red tape in the relationship between government and German business is as old as government control of German business. But as controls expanded, red tape expanded in snow-ball fashion. The resulting condition is what the Germans delight in calling "paper war," namely, a relentless exchange of documents among the various control agencies as well as between each of them and the businessman. By way of illustration, as early as 1937 a case was reported publicly in which an export order involving only some hundreds of RM required action by five different Commodity Control Boards, and more than a hundred letters were written.[102] When, later on, at the time of a visit of Mr. Funk to Bremen, the exporters in that city pasted the walls of a room with specimens of the various forms required for the carrying out of export transactions, the wallpaper was said to have disappeared completely.[103] Were these extreme cases? As late as the spring of 1943, it was revealed that to obtain permission for the construction of a large industrial plant, applications had to be made to about thirty government bureaus.[104] Almost every number of Nazi economic reviews and many issues of German dailies carry stories illustrating what Hamlet called "the insolence of office." The bureaucratic handling of control of business is a perennial subject of Nazi worry. It is freely discussed, universally cursed, and obviously hard to overcome. Every few months, new drives are

[102] *Wirtschaftskurve* II (1937), p. 124.
[103] *Die Deutsche Volkswirtschaft* No. 16 (1943), p. 499.
[104] "Die Baubehoerde," *Frankfurter Zeitung*, Mar. 4, 1943.

started to cure the evil and the very repetition of these drives indicates their failure, or at least merely temporary success.

To many a German and foreign businessman who had taken it for granted that he would get much speedier action from an authoritarian Nazi regime than from a democratic government, the entanglement in red tape has come as a bitter disappointment. Was not promptness in administration a promise held out by the authoritarian principle? Surely a "dynamic" Nazi government would go straight to the core of the evil, ruthlessly cutting out every obstacle to action. It is true the Nazis eliminated every check and balance from the legislative process. Unhampered by an articulate public opinion or by organized pressure groups and without restrictions imposed either by parliament or by the courts, the government issues laws, decrees, rules, and regulations whenever it so desires. In this field there are no delays; the authoritarian principle is being successfully carried to its ultimate limits. But although banned from the legislative process, the checks and balances have not disappeared. Unexpectedly, circuitously, against the will of the government, they have emerged again—in the administrative process. This—it would seem—was due to the extent to which the Nazis have carried the regulation of business rather than to lack of resolve or to incompetence on the part of the German civil service or the business organizations. Obviously, the field of administration has become too inclusive and the administrative process too involved for the authoritarian principle to be effective. Dictatorship or no dictatorship, it is one thing for a government to issue rules and regulations, but to ensure their smooth application is a different matter.

XIV. Summary and Conclusions

We are now in a position to view Nazi control of business as a whole. One feature repeats itself throughout, and stands out above all others:

In every essential respect Nazi control of business is total.

First, control is total as to the phases of business life and operation covered. Each major phase—financial as well as non-financial—is regulated, from the first to the last. German business enterprise is born, is operated, and dies by the will of the government. It is controlled "from the cradle to the grave."

Second, control is total with respect to the units of business enterprise covered. While our analysis has proceeded by phases of life and operation rather than by branches of business, it is clear that with varying techniques control extends to every branch and thereby to every individual German enterprise, from the largest farm to the smallest tenant, from the foremost industrial concern to the humblest cobbler.

Third, control is total in terms of the power and discretion granted the agencies of control. In no case is the government hampered by rights of citizens or firms in going to the furthest length of regulation. Practically every decree authorizing government action to control business includes a provision that action pursuant to the decree shall not be subject to court review. Usually, before a decision is rendered, the German businessman can claim only an informal hearing; afterwards he can appeal only to another agency higher up the ladder—if he can appeal at all! Nor may the citizen or firm claim compensation in the courts for losses or damages sustained as a consequence of action by or on behalf of the controlling agencies. Under Nazi rule, government control of business is not subject to due process of law. There are no legal restrictions on the power of the controlling agencies.

In view of this totality, Nazi control of business cannot be identified with any pattern of control of business (or the equivalent of what we call business) known to have existed in the past. There are elements of feudal control; for example, the worker is tied to his job. The Nazis, however, have out-feudalized the Middle Ages, for they have tied the landlord to his estate and the businessman to his desk. There are elements of corporative control; organization of business on trade lines has been made compulsory. But the Nazis have out-organized both the corporations of late antiquity and the medieval guilds by including agriculture, industry, mining, and transportation —fields which neither corporations nor guilds had ever entered. There are elements of mercantilistic control; for example, prices are fixed, and the government promotes certain industries. But the Nazis have outdone the absolute kings, who fixed prices of only a limited number of commodities, and did not claim control over the supply of capital, raw material, labor, and the allocation of profits, to say nothing of bookkeeping and cost accounting and the other phases of business operation analyzed in this study. There are elements that are usually associated with the condition of slavery; in matters pertaining to the conduct of his affairs, the businessman has been completely subjected to the will of a master. But the Nazis have out-distanced antiquity by substituting one state control for control by scattered private individuals, thereby guaranteeing unity of purpose.

In short, the Nazi regime has borrowed from and revitalized the most important concepts of control found in the annals of history. But it has gone far beyond every one of them. It has created something new. It has built up, has spread abroad, and successfully operates a system of regimentation of business, in scope and intensity such as the world has never seen.

With control of business total, does it make sense to refer to German business as being conducted under a system of private enterprise? Has not the quantity of government controls wrought a fundamental change in the quality of the German economic system?

Total regulation has revolutionized
the character of German business.

It is true—and we have emphasized the fact—that, as a matter of policy, the Nazis have not nationalized business; the bulk of enterprise in Germany continues to be privately owned and operated. However, it is apparent that private ownership and operation of business under the Nazi system of total regulation is not private enterprise as the term is generally understood.

Private ownership undoubtedly yields income, but income so derived has been increasingly absorbed or canalized by the state. As to private operation, all major decisions required for the conduct of business enterprise are imposed by, or made within the framework of, government directives. What can the German businessman do in his capacity as a manager? His hands are either tied or manipulated like marionettes by other hands—the hands of the various government agencies and of the business organizations acting for and under the government. Viewed from below he gives orders; viewed from above he receives orders. Viewed from below he is still the boss; viewed from above he is a flunky. Viewed from below he is a businessman; viewed from above he is a government agent.

In effect, private enterprise has been eclipsed; it is neither private nor enterprise. The formality of extensive government ownership has been avoided, but the means of production have been controlled as thoroughly and as effectively as if they had been owned and operated by the government. Total regulation has done a job equivalent to nationalizing. It was not necessary for the Nazis to convert business into a government department.

The choice of regulation as a method had a distinct advantage as a political strategy. Regulating rather than nationalizing business enterprise, it was possible for the Nazis to preserve private ownership of the means of production as an institution, while in reality reducing it to a meaningless legal concept. Regulating rather than nationalizing business enterprise, it was possible for them to pay lip service to private ownership as a pillar and cornerstone of society, while in reality they were stripping it of all its essential attributes. The Nazis have kept the businessman on the job, and the old civil, commercial, and industrial codes on the statute books. Ostensibly, they were conservative; before a world haunted by the fear of communism they could pose as a bulwark of law and order. Actually, they accomplished a revolution. Henry George was referring not only to the past when he wrote: "It is an axiom of statesmanship which the successful founders of tyranny have understood and acted upon—that great changes can best be brought about under old forms."